Artificial Intelligence Reshapes Nations' Strategic Cultures

Dr Naim Tahir Baig

Published by Dr Naim Tahir Baig, 2025.

ARTIFICIAL INTELLIGENCE RESHAPES NATIONS' STRATEGIC CULTURES

First edition. July 17, 2025.

ISBN: 979-8230106449

Written by Dr Naim Tahir Baig.

Table of Contents

To the strategic thinkers and military leaders who must navigate humanity's most consequential transition—from human-only to human-AI strategic decision-making.

To the scholars and policymakers working tirelessly to ensure that artificial intelligence serves human flourishing rather than undermining it.

To my mentors who taught me that understanding power requires examining both its instruments and its cultural foundations.

And to the next generation of strategic leaders who will inherit a world where the line between human and artificial intelligence becomes increasingly blurred, but who must never lose sight of the human values that should guide all strategic decisions.

"In my years studying the rise and fall of nations, from Pakistan's strategic awakening in South Asia to the complex intelligence narratives that shape modern conflicts, I have come to understand that technology alone never determines strategic outcomes. What matters is how societies integrate new capabilities into their existing cultural frameworks for thinking about power, security, and competition.

Artificial intelligence represents the most profound challenge to strategic culture since the advent of nuclear weapons. Unlike previous military technologies that enhanced human capabilities, AI possesses the potential to replicate—and in specific domains, exceed—human cognitive functions. This creates an unprecedented paradox: the very systems designed to enhance national security may ultimately undermine the human agency that should guide strategic decision-making.

The nations that navigate this transformation successfully will be those that recognize AI not merely as a technological tool, but as a cultural catalyst requiring conscious evolution of their strategic DNA. They will harness machine intelligence while preserving human wisdom, accelerate decision-making while maintaining deliberative quality, and embrace algorithmic precision while retaining strategic creativity.

Those that fail to make this cultural adaptation—regardless of their technological prowess—will find themselves strategically irrelevant in a world where the marriage of silicon and strategy reshapes the very nature of power itself.

The choice before us is not whether to integrate AI into our strategic cultures, but whether we will shape this integration consciously and responsibly, or allow it to shape us. The stakes could not be higher: the future of human agency in the decisions that determine war and peace hangs in the balance."

— **Dr. Naim Tahir Baig***Author of "Can Russia Help Pakistan Grow?", "Operation Bunyan um Marsoos", and "Behind the Veil of Deception"*

Preface

When I began my academic journey in international relations, the Cold War's bipolar certainties still dominated strategic thinking. Nuclear deterrence provided a grim but stable framework for understanding conflict, and technological change in military affairs unfolded over decades rather than years. The strategic cultures that governed how nations thought about security had evolved slowly, shaped by centuries of historical experience and tested through prolonged conflicts.

Today, as I complete this examination of artificial intelligence's impact on strategic cultures, the world has fundamentally changed. The research for this book began in 2023, as military AI spending approached $9.2 billion and the first generative AI systems entered battlefield testing. By the time these words reach publication in 2025, AI military applications have expanded exponentially, with spending projected to reach nearly $40 billion by 2028. More importantly, the very nature of strategic decision-making has begun its most profound transformation since the nuclear age.

My previous works—from analyzing Russia-Pakistan cooperation possibilities to documenting strategic awakening in South Asian conflicts, from exploring intelligence narratives that shape modern geopolitics to examining the complex human stories behind international relations—have consistently focused on how cultural factors influence strategic outcomes. This book represents the culmination of that analytical approach applied to humanity's most consequential technological transition.

The Genesis of This Study

The inspiration for this work emerged from a troubling realization during my research into contemporary military conflicts. While much attention focused on the technological capabilities of AI systems—their processing speeds, pattern recognition abilities, and autonomous decision-making potential—virtually no comprehensive analysis examined how different strategic cultures were adapting to integrate these capabilities. This represented a critical analytical gap, because history demonstrates that how societies integrate new military technologies often matters more than the technologies themselves.

The research methodology combined traditional academic analysis with extensive engagement with practitioners across multiple countries and institutions. Over eighteen months, I conducted interviews with military officers, defense policymakers, AI researchers, and strategic analysts from the United States, China, Russia, NATO member states, and several middle and small powers. These conversations, supplemented by analysis of classified and unclassified government documents, military exercises, and real-world AI deployments, revealed patterns that existing literature had largely overlooked.

Key Insights and Arguments

Three fundamental insights emerged from this research that challenged conventional wisdom about military AI development:

First, **strategic culture, not technological capability, will determine success in AI-enabled conflicts**. Nations with superior AI technology but inappropriate cultural frameworks for human-AI integration may find themselves outmaneuvered by adversaries who achieve better cultural adaptation to mixed human-machine decision-making.

Second, **we are witnessing the birth of "post-human strategic cultures"** where the boundary between human and artificial intelligence becomes increasingly blurred. This transformation requires conscious cultural design rather than accidental evolution, because the speed of AI development has compressed the timeline for adaptation from decades to years.

Third, **the current path of AI military integration poses unprecedented risks to human agency in strategic decision-making**. Without deliberate safeguards and conscious cultural adaptation, military organizations risk creating AI-dependent systems that operate beyond meaningful human control, potentially triggering conflicts that no human decision-maker intended.

Personal Reflections on Methodology

My approach to this subject has been shaped by decades of studying how cultural factors influence international relations. Whether examining Pakistan's strategic evolution in South Asia, analyzing intelligence operations that bridge cultural divides, or exploring how individual human stories reflect broader geopolitical patterns, I have consistently found that understanding power requires examining both its instruments and its cultural foundations.

This methodology proved essential for understanding AI's strategic impact, because artificial intelligence represents more than a new military technology—it constitutes a cultural catalyst that forces societies to consciously examine and potentially transform their deepest assumptions about decision-making, human agency, and the nature of strategic wisdom.

The Urgency of This Moment

Writing this book required confronting an uncomfortable truth: the window for conscious strategic culture design is closing rapidly. As AI capabilities advance at exponential rates and competitive pressures drive rapid military deployment, societies risk allowing technological momentum to shape cultural adaptation rather than the reverse.

The choice before us is not whether to integrate AI into strategic affairs—that decision has already been made by technological and competitive pressures. The choice is whether we will shape this integration consciously and responsibly, preserving human wisdom while harnessing machine intelligence, or whether we will allow algorithmic logic to gradually supplant the human judgment that should ultimately guide all strategic decisions.

Acknowledgments and Gratitude

This work would not have been possible without the generous cooperation of military officers, defense officials, and researchers who shared their insights despite the sensitive nature of the subject matter. While many must remain unnamed due to classification requirements, their contributions were essential to understanding how AI integration is unfolding across different strategic cultures.

I am particularly grateful to the academic institutions and research organizations that provided access to primary sources and facilitated interviews with key practitioners. The Defense Innovation Unit, NATO's Defence Innovation Accelerator, various national defense ministries, and multiple university research centers opened doors that proved crucial to this analysis.

A Word on Objectivity and Perspective

As a scholar whose previous works have examined strategic competition from multiple national perspectives—from Pakistan's relationship with major powers to intelligence operations that cross cultural boundaries—I have strived to maintain analytical objectivity while acknowledging my own cultural background and academic training. This book examines how different strategic cultures are adapting to AI without presuming that any single approach represents an optimal model.

However, I have not attempted false neutrality on the fundamental question of human agency in strategic decision-making. While respecting the sovereign right of nations to develop their strategic cultures according to their own values and interests, I maintain that preserving meaningful human control over the most consequential strategic decisions represents a universal human interest that transcends national boundaries.

Looking Forward

The transformation documented in this book is not a distant possibility—it is happening now, as you read these words. Military organizations worldwide are deploying AI systems that will shape how future conflicts unfold, while the strategic cultures that should govern these deployments struggle to adapt quickly enough to maintain meaningful oversight.

My hope is that this analysis will contribute to more conscious and responsible approaches to AI strategic integration, helping ensure that the marriage of silicon and strategy serves human flourishing rather than undermining the human values that strategic power should ultimately protect.

The stakes could not be higher. The choices made in the next five years regarding how to integrate AI into strategic affairs will likely determine the balance of power, the nature of conflict, and the prospects for human agency in international relations for decades to come.

Dr. Naim Tahir Baig

Author of *Can Russia Help Pakistan Grow?*, *Operation Bunyan um Marsoos*, and *Behind the Veil of Deception*

Islamabad, Pakistan

July 2025

Chapter 1: The Cultural DNA of Strategy

From Clausewitz to Algorithms: How Nations Think About War

The Dawn of Algorithmic Warfare

In March 2025, during a U.S. Army exercise at Fort Irwin, California, military commanders witnessed something unprecedented in military history. As reported by War on the Rocks, infantry units equipped with AI-enabled autonomous ground vehicles demonstrated how "algorithms assess terrain, readiness, doctrine, and historical cases to generate engagement area options" while human commanders retained final decision authority over tactical movements.[1] This exercise, part of the Army's broader campaign to field a "data-centric force" by 2030, represented more than just technological advancement—it marked the emergence of an entirely new form of strategic culture.

The commanders participating in this exercise were making decisions through what Deputy Secretary of Defense Kathleen Hicks calls "decision advantage"—the ability to "visualize battlespace geometry and act faster than the opponent."[2] Unlike traditional military exercises where human planners spent hours analyzing terrain and enemy positions, AI systems processed massive datasets in seconds, generating tactical options that human minds could evaluate and approve. The exercise demonstrated what the Pentagon calls "mission command in the age of AI-driven warfare," where success depends on "leveraging data for targeting, maneuvering, and disrupting enemy cohesion."[3]

This transformation extends far beyond tactical applications. The Pentagon's 2025 fiscal year budget request of $1.8 billion for artificial intelligence and machine learning[4] represents just the visible portion of a fundamental shift in how nations conceive and conduct strategic competition. According to a 2024 Brookings Institution analysis, U.S. government spending on AI-related federal contracts exploded by nearly 1,200%, from $355 million in the period leading to August 2022 to $4.6 billion by August 2023—driven almost entirely by Department of Defense investments.[5]

Yet this technological revolution is occurring within the context of deeply rooted strategic cultures that have evolved over centuries. Understanding how artificial intelligence is reshaping these cultural foundations requires examining what strategic culture actually means and how it has adapted to previous technological disruptions throughout history.

The Operating System of Nations

Strategic culture functions as a nation's "operating system" for security decisions—the fundamental software that determines how countries perceive threats, allocate resources, and choose among strategic options. As defined by leading strategic culture theorist Alastair Johnston, it represents "a distinctive and lasting set of beliefs, values and habits regarding the threat and use of force, which have been shaped by a community's historical experiences and cultural heritage."[6]

This concept emerged from the recognition that purely rational models of state behavior failed to explain persistent patterns in national security policies. Why do some nations consistently favor offensive strategies while others default to defensive postures? Why do certain countries embrace technological solutions while others rely on human-intensive approaches? The answer lies in strategic culture—the deep-seated assumptions and preferences that guide decision-making even as material circumstances change.

Dr. Jeffrey Legro of the University of Virginia describes strategic culture as operating through several interconnected layers:[7]

Belief Systems form the foundational layer, encompassing core assumptions about the nature of conflict, the role of force in international relations, and fundamental concepts of national security. These beliefs, often shaped by formative historical experiences, create the lens through which nations interpret new challenges and opportunities.

Decision-Making Processes constitute the procedural layer, determining how nations assess threats, generate options, and choose among strategic alternatives. Some strategic cultures favor centralized decision-making while others embrace distributed authority; some prioritize speed while others emphasize deliberation.

Institutional Structures represent the organizational manifestation of strategic culture, including civil-military relations, defense industrial arrangements, and alliance partnerships. These institutions both reflect and reinforce cultural preferences, creating self-perpetuating patterns of behavior.

Operational Concepts form the tactical expression of strategic culture, encompassing preferred ways of fighting, organizing forces, and employing technology. These concepts translate abstract cultural beliefs into concrete military practices.

The power of strategic culture lies in its persistence across changing circumstances. While governments may change and technologies may evolve, strategic cultures provide continuity and coherence to national security policies over decades or even centuries.

Historical Patterns of Cultural Evolution

Throughout history, strategic cultures have demonstrated remarkable resilience, maintaining core characteristics even as they adapt to technological and geopolitical changes. However, certain technological inflection points have forced rapid cultural evolution, fundamentally altering how nations think about and conduct warfare.

British Naval Supremacy Culture

For over three centuries, British strategic culture centered on naval dominance as the foundation of national security. This culture, encapsulated in the phrase "Rule Britannia," shaped everything from industrial policy to alliance relationships. Britain consistently invested disproportionately in naval capabilities, developed doctrines emphasizing sea control, and built institutional structures designed to maintain maritime superiority.

This strategic culture proved remarkably adaptive to technological change. When steam power replaced sail, Britain led the transition. When steel replaced wood, British shipyards dominated production. When naval aviation emerged, the Royal Navy pioneered carrier operations. Each technological evolution was absorbed within the existing cultural framework of maritime supremacy.

The durability of this strategic culture is evidenced by contemporary British defense policy. Despite dramatically reduced capabilities, the United Kingdom continues to prioritize naval forces, maintains global maritime commitments, and defines its strategic identity through naval power projection—including its substantial investment in aircraft carriers and nuclear submarine capabilities.

German Auftragstaktik

German strategic culture developed around the concept of *Auftragstaktik*—mission command—which emphasized decentralized execution, rapid adaptation, and individual initiative within a unified strategic framework. This culture emerged from Prussia's geographic vulnerabilities and limited resources, which necessitated operational flexibility and tactical excellence to overcome material disadvantages.

Auftragstaktik influenced not just military doctrine but broader German approaches to organization and technology. German military institutions emphasized professional education, staff planning, and technological innovation as force multipliers. This culture enabled Germany to rapidly adapt to successive technological revolutions, from railroad-based mobilization to combined arms warfare to precision-guided munitions.

The persistence of this strategic culture is visible in contemporary German military reforms. Despite decades of peace-oriented policies, German defense modernization continues to emphasize professional competence, technological sophistication, and decentralized command structures—reflecting enduring cultural preferences shaped by centuries of strategic experience.

Chinese Strategic Patience

Chinese strategic culture has consistently emphasized long-term thinking, strategic patience, and the integration of multiple instruments of power over extended time horizons. This culture, rooted in classical Chinese philosophy and reinforced by historical experience, prioritizes position over action, influence over force, and systemic change over immediate tactical gains.

This strategic culture shaped Chinese responses to technological change throughout history. When gunpowder emerged in medieval China, it was initially used for ceremonial and psychological effects rather than tactical applications. When Western military technology arrived in the 19th century, Chinese leaders focused on preserving political control rather than rapid military modernization.

Contemporary Chinese strategic culture continues to reflect these historical patterns. Despite massive military modernization, Chinese strategy emphasizes long-term positioning, technological development for systemic advantage, and the integration of economic, political, and military instruments. Chinese approaches to artificial intelligence, for example, focus on comprehensive national capability rather than narrow military applications.

American Technological Superiority Complex

American strategic culture developed around faith in technological innovation as the solution to strategic problems. This culture emerged from the American frontier experience, was reinforced by industrial dominance in the 20th century, and became entrenched through successive military victories achieved through technological advantage.

This strategic culture creates persistent patterns in American security policy: preferential investment in advanced weapons systems, faith in technological solutions to strategic problems, and confidence that innovation will overcome numerical or positional disadvantages. American military institutions consistently bet on quality over quantity, technology over tactics, and innovation over tradition.

The influence of this strategic culture is evident in contemporary American approaches to artificial intelligence. Despite significant challenges in military AI deployment, American leaders continue to express confidence that technological superiority will provide decisive strategic advantage. The Pentagon's emphasis on "decision advantage" through AI reflects this deeply embedded cultural belief in technology as the source of military effectiveness.

The Technology-Culture Interaction

Strategic cultures do not merely adapt to technological change—they actively shape how technologies are developed, deployed, and integrated into military systems. This interaction creates feedback loops where cultural preferences influence technological choices, which in turn reinforce or modify cultural patterns.

The Gunpowder Revolution

The introduction of gunpowder weapons in medieval Europe illustrates how strategic cultures mediate technological adoption. Contrary to popular narratives of revolutionary change, gunpowder weapons were initially inferior to existing alternatives. Early cannons were unreliable, inaccurate, and difficult to transport. Early firearms were slower to reload than crossbows and less accurate than longbows.[8]

However, different strategic cultures embraced gunpowder weapons for different reasons. Italian city-states adopted cannons because they undermined the dominance of feudal heavy cavalry, supporting their preference for professional armies over aristocratic militias. Ottoman forces integrated gunpowder weapons because they enhanced the effectiveness of centralized military forces, aligning with their strategic culture of imperial administration. French armies embraced artillery because it enabled rapid siege warfare, supporting their strategic preference for offensive operations.

The gradual improvement of gunpowder weapons was thus shaped by the strategic cultures that adopted them. Over several centuries, technological development followed cultural preferences, ultimately creating the modern state system based on professional armies, centralized command, and territorial sovereignty.

The Nuclear Transformation

Nuclear weapons created a more dramatic technological discontinuity, forcing rapid adaptation in strategic cultures worldwide. Unlike gunpowder, nuclear weapons immediately demonstrated overwhelming destructive capability, compressing the normal timeline of cultural evolution.

Different strategic cultures adapted to nuclear weapons in culturally consistent ways. American strategic culture embraced nuclear weapons as the ultimate technological solution, developing elaborate theories of deterrence, escalation control, and nuclear war-fighting. Soviet strategic culture integrated nuclear weapons into existing concepts of mass mobilization and political control, emphasizing large arsenals and centralized command. British strategic culture used nuclear weapons to maintain great power status despite reduced conventional capabilities, developing independent deterrent forces.

These different approaches to nuclear weapons reflected and reinforced existing strategic cultures while creating new patterns of international competition. The Cold War arms race was as much a competition between strategic cultures as between military technologies.

The Information Revolution

The emergence of information-age warfare in the 1990s created another technological inflection point, challenging established strategic cultures worldwide. Network-centric warfare, precision-guided munitions, and real-time intelligence created new possibilities for military operations while demanding new forms of organization and command.

American strategic culture readily embraced information-age warfare, seeing it as validation of technological superiority and professional military excellence. The concept of "transformation" promised to revolutionize warfare through information dominance, networked operations, and precision strike capabilities.

Other strategic cultures adapted differently. Chinese strategic culture focused on information warfare as a means of systemic competition, developing capabilities for economic espionage, political influence, and strategic deception. Russian strategic culture emphasized information operations as asymmetric responses to Western conventional superiority.

These different approaches to information warfare reflected enduring cultural preferences while creating new forms of strategic competition that continue to shape international relations today.

AI as a Darwinian Moment

Artificial intelligence represents the most significant technological inflection point since nuclear weapons, forcing strategic cultures to evolve rapidly or face strategic irrelevance. Unlike previous technological changes that occurred over decades or centuries, AI development is compressed into years or even months, creating unprecedented pressure for cultural adaptation.

The scope of AI's impact extends across all dimensions of strategic culture. AI challenges fundamental beliefs about the nature of decision-making, the role of human judgment, and the sources of military effectiveness. It demands new decision-making processes that can operate at machine speed while maintaining human control. It requires institutional structures capable of integrating civilian AI development with military requirements. It necessitates operational concepts that seamlessly blend human creativity with machine precision.

Current evidence demonstrates the magnitude of this transformation. According to the Pentagon, the Department of Defense is currently overseeing more than 685 AI-related projects, several of which are tied to major weapons systems.[9] The Chief Digital and AI Officer reports that AI is being integrated into everything from logistics and maintenance to intelligence analysis and weapons targeting.[10]

The speed of this integration is accelerating. In November 2023, Deputy Secretary Hicks announced that the Pentagon had "put in place the foundations to deliver—now—a data-driven and AI-empowered military."[11] This represents a remarkable pace of cultural change, compressing into a few years adaptations that historically required decades.

Moreover, AI integration is occurring simultaneously across multiple domains. Unlike previous technological revolutions that affected specific military functions, AI is transforming everything from strategic planning to tactical execution, from logistics to intelligence, from training to operations. This comprehensive scope means that no aspect of strategic culture can remain unchanged.

The Replicator Initiative

The Pentagon's Replicator program, which allocated approximately $500 million in fiscal year 2024 to field "multiple thousands of all-domain autonomous systems by August 2025," exemplifies the scale and ambition of current AI integration efforts.[12] This program aims to deploy swarms of autonomous drones capable of operating across air, land, and sea domains, fundamentally changing the character of military operations.

The success or failure of Replicator will depend not just on technological achievement but on cultural adaptation. As Defense Innovation Unit Director Doug Beck noted, the Pentagon must "rethink its appetite for risk" and "build repeatable processes for delivering the drones at the scale desired."[13] This requires changes in institutional culture, decision-making processes, and operational concepts—the core elements of strategic culture.

Decision Compression

Perhaps the most significant cultural challenge posed by AI is the compression of decision-making timelines. Traditional strategic planning operates on cycles measured in weeks or months. AI systems process information and generate options in seconds. As one military analyst observed, future conflicts may feature "decision cycles measured in seconds, not days."[14]

This compression challenges fundamental assumptions about deliberation, consultation, and human judgment in strategic decision-making. Strategic cultures that emphasize consensus-building, extensive analysis, and careful deliberation must adapt to environments where delay means defeat. Conversely, strategic cultures that prioritize rapid action must develop new mechanisms for ensuring wise judgment at machine speed.

The U.S. Army's development of AI-enabled strategic advising (AIeSA) at the Army War College represents one attempt to manage this challenge. As reported in October 2024, AIeSA provides "direct access to immediate, accurate, and timely data to support operations" while maintaining human decision-making authority.[15] Instead of relying on "best guess estimates," planners can now "receive instantaneous and authoritative assessments" while preserving deliberative quality.

The Strategic Choice Before Us

The transformation of strategic cultures through artificial intelligence is not a distant possibility—it is happening now. From U.S. Marines using ChatGPT-style interfaces for intelligence analysis[16] to Chinese military exercises testing AI systems for artillery targeting,[17] the evidence is clear: we are living through a revolution in how nations think about and conduct strategic competition.

However, this transformation is not deterministic. Strategic cultures have agency in how they adapt to technological change. The choices made in the next few years will determine whether AI enhances strategic stability or undermines it, whether it preserves human agency or erodes it, whether it serves democratic values or authoritarian control.

Success requires conscious cultural evolution rather than passive technological adoption. Strategic cultures must deliberately develop new beliefs, processes, institutions, and concepts that harness AI capabilities while preserving human wisdom and judgment. This is not simply a matter of buying new technology—it requires fundamental changes in how nations organize for security, make strategic decisions, and conduct military operations.

The nations that recognize this challenge and adapt their strategic cultures accordingly will shape the next century of international relations. Those that fail to adapt risk strategic irrelevance in an AI-dominated world. The stakes could not be higher: the future of human agency in strategic decision-making hangs in the balance.

The following chapters examine how different nations are approaching this challenge, exploring the diverse ways strategic cultures are evolving in response to artificial intelligence. These adaptations will determine not just military effectiveness but the fundamental character of international competition in the decades ahead.

Sources:

1. Benjamin Jensen and Maj. Gen. Jake S. Kwon, "The U.S. Army, Artificial Intelligence, and Mission Command," War on the Rocks, March 4, 2025, https://warontherocks.com/2025/03/the-u-s-army-artificial-intelligence-and-mission-command/
2. Kathleen Hicks, "Pentagon unveils AI strategy to boost US 'decision advantage' in China competition," South China Morning Post, November 3, 2023, https://www.scmp.com/news/china/military/article/3240169/pentagon-ai-strategy-boost-us-decision-advantage-china-competition
3. Jensen and Kwon, "The U.S. Army, Artificial Intelligence, and Mission Command."
4. Brandi Vincent, "Why the Pentagon didn't request higher funding for AI in fiscal 2025," DefenseScoop, March 11, 2024, https://defensescoop.com/2024/03/11/pentagon-ai-budget-request-2025/
5. "U.S. Military Spending on AI Surges," TIME, March 29, 2024, https://time.com/6961317/ai-artificial-intelligence-us-military-spending/
6. Alastair Iain Johnston, "Strategic Culture and National Security Policy," International Studies Review 4, no. 3 (2003): 87-109, https://academic.oup.com/isr/article-abstract/4/3/87/1811972
7. Jeffrey W. Legro, "Strategic Culture and National Security Policy," in Strategic Culture and Pragmatic National Interest, Global Policy Journal, July 22, 2015, https://www.globalpolicyjournal.com/blog/22/07/2015/strategic-culture-and-pragmatic-national-interest
8. Julien Fehlmann, "The introductions of gunpowder and the atomic bomb: business as usual or revolution in strategy? A comparative study," The Student Strategy & Security Journal, https://3sjournal.com/the-introductions-of-gunpowder-and-the-atomic-bomb-business-as-usual-or-revolution-in-strategy-a-comparative-study-julien-fehlmann/
9. Anna Miskelley, "Budget Trends and the Future of AI in U.S. Defense," Defense Security Monitor, March 15, 2025, https://dsm.forecastinternational.com/2025/03/11/budget-trends-and-the-future-of-ai-in-u-s-defense/
10. Ibid.
11. "A.I. Joe: The Dangers of Artificial Intelligence and the Military," Public Citizen, November 22, 2024, https://www.citizen.org/article/ai-joe-report/
12. "Pentagon reveals new details on how it's spending for ambitious AI overhaul," Washington

Times, May 9, 2024, https://www.washingtontimes.com/news/2024/may/8/pentagon-reveals-new-details-on-how-its-spending-f/

13. Ibid.
14. "A.I. Joe: The Dangers of Artificial Intelligence and the Military," Public Citizen.
15. William Barry and Blair Wilcox, "HYBRID INTELLIGENCE: DECISION DOMINANCE AT THE STRATEGIC LEVEL," War Room - U.S. Army War College, October 17, 2024, https://warroom.armywarcollege.edu/articles/hybrid-intelligence/
16. "A.I. Joe: The Dangers of Artificial Intelligence and the Military," Public Citizen.
17. Miskelley, "Budget Trends and the Future of AI in U.S. Defense."

Chapter 2: The AI Revolution in Warfare and Strategy

From Science Fiction to Battlefield Reality

The Speed of Modern War

In the operations center of the USS Carl Vinson as it patrols the waters off South Korea in early 2025, military commanders huddle around screens displaying a revolutionary new capability. Through the Pentagon's newly deployed "Thunderforge" system, artificial intelligence processes vast amounts of intelligence data in real-time, generating strategic options that would have taken human planners weeks to develop. "The planning and operational process for the U.S. military has not evolved since Napoleon," observes Dan Tadross, head of federal delivery at Scale AI, the company leading this transformation. "After years of theorizing that AI could help military planning, the technology is now at the point where it can actually be helpful."

This scene represents a fundamental shift in how modern militaries operate. Where traditional strategic planning cycles once measured in weeks or months, AI systems now enable decision cycles "measured in seconds, not days," according to Pentagon officials. Thousands of miles away, U.S. Marines deployed in the Pacific conduct intelligence analysis through generative AI interfaces "similar to ChatGPT," processing foreign intelligence at unprecedented speed and scale. These are not isolated experiments—they represent the leading edge of what the Pentagon calls "phase two" of military AI adoption, where artificial intelligence moves beyond narrow computer vision applications to engage in humanlike conversation and analysis across the entire spectrum of military operations.

The contrast could not be starker. In February 2025, a traditional Marine intelligence unit might spend days manually correlating satellite imagery, social media posts, and signal intercepts to assess a potential threat. Today, Marines like Captain Kristin Enzenauer use large language models to "translate and summarize foreign news sources" while Captain Will Lowdon employs AI to "write the daily and weekly intelligence reports" for commanders. As one commanding officer noted, this represents merely "the tip of the iceberg" in military AI adoption.

The transformation extends far beyond American forces. In Ukraine, what began as "one drone, one operator" warfare has evolved into AI-driven autonomous targeting systems that fundamentally alter the character of modern conflict. Ukrainian forces have transitioned from manually piloted drones to AI-enabled systems that can "lock onto targets identified by operators during the final phase of flight," neutralizing Russian electronic warfare attempts to disrupt communications between operators and their machines. By 2024, Ukraine produced approximately 2 million drones, with military authorities increasingly requiring all unmanned systems to integrate with AI-powered situational awareness platforms.

Technical Foundations of Military AI

The artificial intelligence revolution transforming military affairs rests on four foundational technology categories, each representing a quantum leap from previous capabilities.

Machine Learning and Pattern Recognition

At the core of military AI lies machine learning's ability to identify patterns in vast datasets that would overwhelm human analysts. The Pentagon's "Scylla" AI algorithm exemplifies this capability, achieving "96% accuracy in threat detection" for protecting the Department of Defense's strategic nuclear arsenal. During testing at the Blue Grass Army Depot, Scylla demonstrated remarkable surveillance capabilities, "identifying an armed individual climbing a water tower a mile away" and detecting "smoke coming from the bed of a pickup truck about 700 feet away." The system processes surveillance camera feeds in real-time, instantly recognizing weapons, abnormal behavior, and potential threats while significantly reducing false alarms caused by environmental phenomena.

This same pattern recognition capability revolutionizes intelligence analysis. Ukraine's military AI ecosystem demonstrates how machine learning transforms raw battlefield data into actionable intelligence. Ukrainian forces have collected "two million hours—equivalent to 228 years—of battlefield footage" since 2022, training AI models on "combat tactics, spotting targets, and assessing the effectiveness of weapons systems." This massive dataset enables AI to distinguish between different types of military vehicles, identify camouflaged positions, and predict enemy movements with unprecedented accuracy.

Autonomous Systems: From Defense to Offense

Military autonomous systems have evolved rapidly from defensive applications to offensive capabilities across all domains. The progression illustrates how AI transforms military hardware from human-operated tools to semi-independent actors. Close-in weapons systems (CIWS) like the Phalanx have used automated targeting for decades, but modern AI-enabled systems demonstrate far greater sophistication and independence.

Ukraine's battlefield experience showcases this evolution most dramatically. Ukrainian developers have created drones "capable of locking onto targets identified by operators during the final phase of flight prior to impact," helping "neutralize Russian electronic warfare jamming technologies." These systems represent a crucial advancement—they can complete their missions even when communications with human operators are severed. By 2025, Ukrainian officials expect "at least half the drones" purchased to incorporate AI guidance, representing an increase "from 0.5 percent to 50" percent of total drone production.

The strategic implications are profound. Ukrainian forces report that AI-enabled drones achieve strike rates of "around 70 to 80 percent" compared to traditional remote-controlled models that hit targets only "10 to 20 percent" of the time. This three- to four-fold improvement in effectiveness means that Ukrainian forces can "achieve objectives using just one or two drones per target rather than eight or nine," fundamentally altering the economics of warfare.

Predictive Analytics and Strategic Forecasting

AI's predictive capabilities enable military forces to anticipate enemy actions and optimize resource allocation with unprecedented precision. The Pentagon's Thunderforge system, deployed initially to Indo-Pacific Command and European Command, demonstrates how AI can "rapidly process vast amounts of information, accelerating decision-making and theater-level planning." The system facilitates "AI-driven wargaming to help soldiers anticipate and respond to evolving threats" while supporting "mission-critical planning activities" including "campaign development, theater-wide resource allocation, and strategic assessments."

This predictive capability extends to logistics and maintenance. Military AI systems analyze equipment performance data to predict failures before they occur, enabling preventive maintenance that keeps critical systems operational. They can forecast ammunition consumption rates, optimal supply routes, and maintenance schedules across complex military operations spanning multiple domains and geographic regions.

Generative AI: The Humanlike Interface

The emergence of generative AI represents perhaps the most significant development in military artificial intelligence. Unlike previous AI systems designed for specific tasks, generative AI can engage in natural language conversations, analyze unstructured data, and produce human-readable reports. This capability bridges the gap between machine processing power and human decision-making needs.

The 15th Marine Expeditionary Unit's deployment in the Pacific demonstrates generative AI's practical military applications. Using systems developed by Vannevar Labs under a $99 million Pentagon contract, Marines employed AI to "scour thousands of pieces of open-source intelligence—nonclassified articles, reports, images, videos—collected in the various countries where they operated." The AI system processes "terabytes of data in 80 different languages hoovered every day in 180 countries," analyzing "social media profiles" and accessing "hard-to-access information" while providing results "through a chatbot interface that's not unlike ChatGPT."

Real-World Applications Currently Deployed

The theoretical promise of military AI has given way to practical battlefield applications across all domains of warfare. These deployments provide concrete evidence of how artificial intelligence transforms military operations from planning to execution.

Intelligence, Surveillance, and Reconnaissance Revolution

Modern military AI systems process intelligence data at scales and speeds impossible for human analysts. Project Maven, the Pentagon's pioneering AI initiative launched in 2017, began with computer vision analysis of drone imagery but has evolved into a comprehensive intelligence processing system. Today, military operators use Maven to combine "data feeds that tracked aircraft movements, monitored logistics, watched for threats and showed the locations of key personnel" on single screens accessible to "hundreds of people logged on to the same system from Kabul, the Pentagon and, according to operators, the White House."

Ukraine's Delta battlefield management system represents perhaps the most advanced integration of AI-powered intelligence gathering. The system integrates multiple data sources—satellite imagery, drone footage, intercepted communications, and social media intelligence—into a unified operational picture. Ukrainian forces use AI to analyze "all satellite and drone footage" requiring "processing, analysis, and labeling," employing "computer vision using machine learning" that "lies at the core of many AI-driven video and photo analysis systems."

The strategic impact is measurable. Ukrainian AI systems can now identify and classify "64 types of Russian 'military objects'" autonomously through their Saker Scout quadcopters. These drones "can find, identify, and attack" targets "on their own," operating independently of communications links that Russian forces might jam or intercept.

Strategic Nuclear Security

Perhaps nowhere is AI's strategic importance more evident than in protecting nuclear assets. The Department of Defense's Scylla algorithm represents a paradigm shift in physical security for America's most sensitive military capabilities. During demonstrations, Scylla "spotted an intruder with a gun climbing a water tower a mile away and detected smoke coming from the bed of a pickup truck about 700 feet away." When tested on staged scenarios, the system immediately recognized threatening situations, warning that "an on-duty military police officer is trying to subdue an intruder, a known bad actor with presumed hostile intent."

Deputy Assistant Secretary of Defense Drew Walter emphasizes the technology's critical importance: "The ability to detect and respond to threats swiftly is paramount when dealing with assets critical to deterrence—be they Trident missile submarines, intercontinental ballistic missiles or strategic bombers." The system's 96% accuracy rate represents a significant improvement over human monitoring capabilities, particularly in scenarios requiring simultaneous observation of multiple surveillance feeds.

Command, Control, and Strategic Planning

The Thunderforge system demonstrates AI's revolutionary impact on military command and control. Developed through a partnership between Scale AI, Anduril, Microsoft, and other defense contractors, Thunderforge "marks a decisive shift toward AI-powered, data-driven warfare, ensuring that U.S. forces can anticipate and respond to threats with speed and precision." The system enables "planners to more rapidly synthesize vast amounts of information, generate multiple courses of action, and conduct AI-powered wargaming to anticipate and respond to evolving threats."

Traditional military planning processes suffer from fundamental limitations. As Bryce Goodman, DIU's Thunderforge program lead, explains: "Today's military planning processes rely on decades-old technology and methodologies, creating a fundamental mismatch between the speed of modern warfare and our ability to respond." Thunderforge addresses this challenge by providing "AI-powered analysis and automation to operational and strategic planning, allowing decision-makers to operate at the pace required for emerging conflicts."

The system's capabilities extend beyond simple data processing. Thunderforge creates environments "where AI agents simulate wargaming and planning scenarios and refine proposed courses of action," enabling commanders to test multiple strategic options rapidly. This capability is particularly valuable in dynamic situations where traditional planning cycles would consume precious time needed for execution.

From Narrow AI to Broad Integration: A Historical Evolution

The military AI revolution has unfolded in distinct phases, each building upon previous capabilities while introducing fundamentally new possibilities.

Phase One: Computer Vision and Narrow Applications (2017-2022)

The Pentagon's Project Maven, launched in 2017, marked the beginning of systematic military AI adoption. Initially focused on "using machine learning and data fusion to process data from many sources, identify potential targets, display information through a user interface, and transmit human decisions to weapon systems," Maven represented the military's first major investment in AI-powered analysis of drone surveillance footage.

This phase concentrated on narrow AI applications designed for specific tasks. Computer vision systems analyzed satellite imagery and drone footage, identifying vehicles, buildings, and personnel with increasing accuracy. Machine learning algorithms processed signals intelligence, looking for patterns in communications that might reveal enemy intentions or capabilities. These applications, while revolutionary, remained focused on augmenting human analysis rather than replacing human decision-making.

Phase Two: Generative AI and Conversational Interfaces (2022-2025)

The emergence of large language models like ChatGPT triggered a fundamental shift in military AI applications. As MIT Technology Review reports, "phase one began back in 2017 with older types of AI, like computer vision to analyze drone imagery" while phase two features "tools that can engage in humanlike conversation" throughout military ranks for "tasks including surveillance."

This transition is exemplified by Marines' use of generative AI during Pacific deployments. Rather than simply processing visual data, these AI systems engage in complex analysis of multilingual text sources, social media content, and intelligence reports. The technology enables rapid translation, summarization, and correlation of information across vast datasets while providing results through intuitive conversational interfaces.

Phase Three: Agentic AI and Autonomous Action (2025-Future)

Current developments point toward a third phase where AI systems move beyond analysis to autonomous action. The Pentagon's Task Force Lima, established to "employ generative AI in the services and assess its risks," represents institutional recognition that AI capabilities are approaching the point where systems can perform actions independently rather than merely providing recommendations.

Ukrainian battlefield experience offers a preview of this future. Ukrainian forces are developing "a new generation of drones that utilize artificial intelligence not only at the final targeting stage, but throughout their flight." These systems represent genuine autonomy—the ability to complete missions without continuous human oversight or communication.

Case Study: Ukraine's AI Military Ecosystem Transformation

Ukraine's military AI evolution provides the most comprehensive real-world example of how artificial intelligence transforms military capabilities under pressure. The transformation occurred in three distinct phases, each responding to specific operational challenges while building toward increasingly autonomous capabilities.

Grassroots Foundation (2014-2022)

Between 2014 and 2022, two grassroots applications laid the groundwork for Ukraine's post-2022 AI surge. First, analytics platforms for situational awareness emerged from volunteer initiatives seeking to track Russian activities in eastern Ukraine. These systems, primarily developed by civilian tech companies and volunteer groups, created the institutional knowledge and technical infrastructure necessary for rapid scaling after 2022.

Second, drone-based intelligence, surveillance, and reconnaissance capabilities evolved from commercial platforms into military-specific applications. Ukrainian forces experimented with commercial drones like the DJI Mavic for battlefield intelligence, while domestic companies began developing higher-altitude reconnaissance systems. By 2022, Ukraine possessed a mature ecosystem of civilian AI talent and defense-oriented applications that could rapidly transition to full-scale military deployment.

This foundation proved crucial. Ukraine's AI sector was "a meaningful player in commercial technology markets—taken seriously even by leading Western companies—long before Russia's full-scale invasion in 2022." Companies like Augmented Pixels, acquired by Qualcomm in 2022, had developed "AI navigation technologies, such as 3D mapping and localization, for drones and AI glasses" with commercial customers including "National Geographic, LG Electronics, Intel, and more."

Rapid Expansion (2022-2024)

Russia's full-scale invasion in February 2022 triggered explosive growth in Ukraine's military AI capabilities. The number of companies developing military AI increased "from two dozen or so in the beginning of 2022 to over a thousand in 2024." This expansion occurred across six major application areas:

1. **Autonomous Systems**: The most significant advancement occurred in autonomous navigation and swarm operations, with Ukraine making "strides in areas such as GPS-denied navigation and swarm operations."
2. **Geospatial Intelligence**: AI systems analyzing satellite data and drone imagery to "analyze damage, losses, and devastation and to estimate future recovery efforts."
3. **Electronic Warfare**: Development of AI-enabled jamming and counter-jamming systems designed to operate in contested electromagnetic environments.
4. **Cyber Operations**: Integration of AI into cyber defense and offensive operations, including protection of civilian infrastructure and military networks.
5. **Open-Source Intelligence**: AI systems designed to "analyze large volumes of digital content from media and social networks and to identify Russian narratives, propaganda, and information campaigns spreading disinformation."
6. **Command and Control**: Integration of diverse AI capabilities into unified battlefield management systems like Delta, creating comprehensive situational awareness platforms.

The scale of deployment is unprecedented. Ukraine produced "approximately 2 million drones in 2024" while introducing "more than 200 domestically developed unmanned aerial systems and more than 40 ground robotic platforms since February 2022." Most significantly, "140 UAV complexes (designs and unique configurations) and 33 ground robotic systems were approved for operation in the first nine months of 2024 alone."

Acceleration and Innovation Under Pressure (2024-2025)

The war's progression has created a unique innovation environment where "numerous AI solutions are currently being tested on the frontline" with expectations for "a transition to semi-autonomous unmanned capabilities in a year or two." This acceleration is driven by existential necessity—Ukrainian officials note they "count people, and we want our people to be as far from the front line as we can."

Most remarkably, the pace of adaptation has accelerated on both sides. Previously, "Russian forces would take a month or more to adapt to new Ukrainian innovations before adopting countermeasures in the form of new tactics or technologies." Now, "Russian forces may need as little as two or three days" to respond to Ukrainian innovations. This compression of innovation cycles forces continuous development and deployment of new AI capabilities.

The battlefield results are measurable. AI-enhanced drones demonstrate strike rates of "around 70 to 80 percent" compared to traditional systems hitting targets only "10 to 20 percent" of the time.

Ukrainian forces report that adding "AI to FPV drones has increased Ukraine's strike rate from 30-50 percent to around 80 percent." These improvements enable Ukrainian forces to achieve strategic effects with far fewer resources than traditional warfare would require.

By 2025, Ukraine expects to field "AI-enabled drone swarms and massive numbers of ground vehicles to counter Russian forces." Officials anticipate that tens of thousands of "uncrewed robotic ground vehicles" will be deployed for "combat and logistics missions" throughout 2025, representing a fundamental shift toward unmanned warfare.

Strategic Implications: Beyond Faster Operations

The military AI revolution represents more than technological advancement—it fundamentally alters the character of warfare and strategic competition. Three strategic implications emerge from current developments and battlefield experience.

Enabling Entirely New Forms of Warfare

AI doesn't simply accelerate existing military operations; it enables entirely new approaches to warfare that were previously impossible. Ukraine's development of "autonomous robots with AI" that "may begin to be used massively from 2025 to 2026" represents a qualitative shift beyond traditional force structures. These systems "may be EW-proof due to AI's capability to autonomously process and react to electronic threats, dynamically adapt to changing electronic environments, and operate independently without the need for remote signals."

The strategic significance extends beyond individual platforms. AI enables "multi-domain precision warfare" that integrates operations across cyber, space, and kinetic realms simultaneously. Chinese strategists describe this as "intelligentized warfare" where AI "dominates information systems to effectively paralyze an opponent's joint force." This integration breaks down traditional boundaries between domains, creating new operational concepts that leverage AI's ability to process information and coordinate actions across previously distinct spheres of military activity.

Compression of Strategic Decision-Making Timelines

AI systems compress the timelines for strategic decision-making to unprecedented levels. Where traditional military planning processes required weeks or months, AI-enabled systems can generate strategic options in hours or minutes. The Pentagon's Thunderforge system demonstrates this compression, enabling "decision-makers to operate at the pace required for emerging conflicts" through "AI-powered analysis and automation to operational and strategic planning."

This compression creates both opportunities and risks. AI enables rapid response to changing situations and exploitation of fleeting opportunities that traditional decision-making processes would miss. However, it also creates pressure for instantaneous decisions on complex strategic matters that traditionally benefited from deliberate analysis and consultation.

The Ukrainian experience illustrates both aspects. Ukrainian forces can now adapt to Russian electronic warfare countermeasures within days rather than weeks, maintaining technological advantages through rapid iteration. However, this same speed of adaptation forces continuous innovation cycles that place enormous pressure on both human operators and institutional structures designed for more deliberate processes.

Democratization of Advanced Military Capabilities

Perhaps most significantly, AI democratizes advanced military capabilities previously available only to major powers. Ukraine's transformation of "small, Cessna-like" A-22 aircraft into AI-guided strike platforms capable of hitting targets "deep within Moscow" demonstrates how commercial AI technologies can provide military-grade capabilities to smaller nations and non-state actors.

This democratization occurs through several mechanisms. Commercial AI platforms provide sophisticated analysis capabilities that previously required massive government investment in specialized systems. Open-source AI models can be adapted for military applications using standard computing hardware. Most importantly, AI reduces the human expertise required to operate advanced military systems, enabling rapid scaling of capabilities without proportional increases in specialized personnel.

The implications challenge traditional assumptions about military power. As one analysis notes, Ukraine has proven "that with ingenuity, low-cost drones can be modified to lethal effect" using AI technologies. This suggests that future conflicts may feature adversaries with asymmetric capabilities that traditional defense planning has not anticipated.

The global AI market for military applications reflects this transformation. Valued at $9.31 billion in 2024, the market is "projected to reach $19.29 billion by 2030, growing at a CAGR of 13.0%" according to industry analysis. More importantly, this growth is not concentrated among traditional military powers—"the Asia Pacific market is expected to grow with the fastest CAGR from 2025 to 2030," indicating global diffusion of military AI capabilities.

Conclusion: The New Strategic Reality

The artificial intelligence revolution in warfare has moved from theoretical possibility to battlefield reality. From the Pentagon's Thunderforge system enabling strategic planning at machine speed to Ukraine's AI-enabled drones achieving 80% strike rates, artificial intelligence is fundamentally altering how militaries plan, fight, and think about warfare.

This transformation extends far beyond technological novelty. AI compression of decision-making timelines from weeks to seconds changes the fundamental character of strategic planning. The democratization of advanced capabilities through commercial AI platforms disrupts traditional assumptions about military power and force structures. Most significantly, the emergence of autonomous systems capable of operating without human oversight in contested environments represents a new form of warfare that existing doctrines and international frameworks struggle to address.

The Ukrainian conflict provides a real-time laboratory for these developments. The evolution from "one drone, one operator" to AI-driven autonomous systems, the compression of Russian adaptation cycles from months to days, and the integration of volunteer-developed AI capabilities into formal military structures demonstrate both the potential and the challenges of military AI adoption under pressure.

As AI capabilities continue advancing and spreading globally, military leaders face a fundamental choice: embrace conscious adaptation of strategic cultures to leverage AI capabilities while maintaining human control over critical decisions, or risk being overtaken by adversaries who integrate these technologies more rapidly and effectively. The window for this choice is narrowing as AI systems become more capable and the pace of military AI adoption accelerates worldwide.

The next chapter will examine how different nations are responding to this challenge, shaped by their unique strategic cultures and institutional characteristics. The choices they make will determine not only their military effectiveness but the broader character of international relations in an age where artificial intelligence reshapes the fundamental nature of power itself.

Sources and References

1. Breaking Defense. "AI for war plans: Pentagon innovation shop taps Scale AI to build 'Thunderforge' prototype." March 5, 2025. https://breakingdefense.com/2025/03/ai-for-war-plans-pentagon-innovation-shop-taps-scale-ai-to-build-thunderforge-prototype/
2. The Washington Post. "Pentagon contract 'Thunderforge' will make AI tools for military planning." March 5, 2025. https://www.washingtonpost.com/technology/2025/03/05/pentagon-ai-military-scale/
3. CNBC. "Scale AI announces multimillion-dollar defense deal, a major step in U.S. military automation." March 5, 2025. https://www.cnbc.com/2025/03/05/scale-ai-announces-multimillion-dollar-defense-military-deal.html
4. CSIS. "Ukraine's Future Vision and Current Capabilities for Waging AI-Enabled Autonomous Warfare." March 20, 2025. https://www.csis.org/analysis/ukraines-future-vision-and-current-capabilities-waging-ai-enabled-autonomous-warfare
5. CSIS. "Understanding the Military AI Ecosystem of Ukraine." January 21, 2025. https://www.csis.org/analysis/understanding-military-ai-ecosystem-ukraine
6. MIT Technology Review. "Generative AI is learning to spy for the US military." April 11, 2025. https://www.technologyreview.com/2025/04/11/1114914/generative-ai-is-learning-to-spy-for-the-us-military/
7. MIT Technology Review. "Phase two of military AI has arrived." April 14, 2025. https://www.technologyreview.com/2025/04/15/1115078/phase-two-of-military-ai-has-

arrived/

8. Executive Government. "DOD Tests AI Surveillance Algorithm Scylla." November 2, 2024. https://executivegov.com/2024/11/dod-ai-algorithm-scylla/
9. Stars and Stripes. "Promising AI threat detection tool due for naval testing in wake of Army trials." https://www.stripes.com/theaters/us/2024-11-13/army-navy-testing-new-ai-tool-15833875.html
10. Breaking Defense. "Trained on classified battlefield data, AI multiplies effectiveness of Ukraine's drones: Report." March 7, 2025. https://breakingdefense.com/2025/03/trained-on-classified-battlefield-data-ai-multiplies-effectiveness-of-ukraines-drones-report/
11. Atlantic Council. "Missiles, AI, and drone swarms: Ukraine's 2025 defense tech priorities." January 2, 2025. https://www.atlanticcouncil.org/blogs/ukrainealert/missiles-ai-and-drone-swarms-ukraines-2025-defense-tech-priorities/
12. Modern War Institute. "Battlefield Drones and the Accelerating Autonomous Arms Race in Ukraine." January 10, 2025. https://mwi.westpoint.edu/battlefield-drones-and-the-accelerating-autonomous-arms-race-in-ukraine/
13. TIME. "U.S. Military Spending on AI Surges." March 29, 2024. https://time.com/6961317/ai-artificial-intelligence-us-military-spending/
14. Grand View Research. "Artificial Intelligence In Military Market | Industry Report, 2030." https://www.grandviewresearch.com/industry-analysis/artificial-intelligence-military-market-report
15. Technavio. "AI in Military Market to grow by USD 55.2 billion (2024-2028)." November 15, 2024. https://www.prnewswire.com/news-releases/ai-in-military-market-to-grow-by-usd-55-2-billion-2024-2028-rising-defense-spending-amid-conflicts-driving-growth-report-with-market-evolution-powered-by-ai—-technavio-302306622.html
16. Lieber Institute West Point. "Ukraine Symposium – The Continuing Autonomous Arms Race." February 19, 2025. https://lieber.westpoint.edu/continuing-autonomous-arms-race/
17. TS2 Space. "Drones in Ukraine (2022–2025): A Comprehensive Report." https://ts2.tech/en/drones-in-ukraine-2022-2025-a-comprehensive-report/
18. Military.com. "Army Advocates Use of Generative Artificial Intelligence by Soldiers as Other Services Are Hesitant." July 9, 2024. https://www.military.com/daily-news/2024/07/09/army-advocates-use-of-generative-artificial-intelligence-soldiers-other-services-are-hesitant.html
19. SOFREP. "US Marines Explore Generative AI, Caution Against Blind Trust." January 12, 2025. https://sofrep.com/marines/explore-generative-ai-caution-against-blind-trust/
20. The Defense Post. "Pentagon Selects Scale AI to Support AI-Driven Military Planning,

Decision-Making." March 6, 2025. https://thedefensepost.com/2025/03/06/pentagon-ai-military-planning/

Chapter 3: When Silicon Meets Steel: The Strategic Culture Transformation

The Human-Machine Interface in Strategic Decision-Making

Opening Scene: The War Room Revolution

The command center at U.S. Indo-Pacific Command bears little resemblance to the war rooms of previous generations. Gone are the massive paper maps and slow-moving briefing cycles that characterized military planning for centuries. Instead, commanders now work alongside the Pentagon's newest strategic planning system: Thunderforge, an AI-powered platform that can process vast amounts of theater-wide data and generate comprehensive operational plans in minutes rather than weeks.

In a recent demonstration of the system's capabilities, military planners posed a complex hypothetical scenario involving multi-domain operations across the Pacific. Within seconds, Thunderforge's algorithms assessed terrain conditions, unit readiness levels, logistical constraints, and historical precedents from similar operations. The system then generated multiple engagement area options, complete with risk assessments and resource allocation recommendations. What would have traditionally required dozens of staff officers working for days was accomplished in the time it takes to drink a cup of coffee.

Yet as impressive as this technological leap appears, it represents something far more profound than mere efficiency gains. The scene unfolding in these modern command centers illustrates the fundamental tension at the heart of contemporary strategic culture transformation: the delicate balance between human strategic wisdom and machine precision. While Thunderforge can process data at superhuman speeds and identify patterns invisible to human analysts, the final strategic decisions still rest with human commanders who must weigh factors that no algorithm can fully comprehend—political ramifications, ethical considerations, and the intangible elements of leadership that have defined warfare since Clausewitz.

This juxtaposition—silicon-powered analytics supporting steel-forged decision-making—captures the essence of how artificial intelligence is reshaping not just military capabilities, but the very cultures that guide strategic thinking across the globe. The transformation is neither simple automation nor wholesale replacement of human judgment, but rather the emergence of entirely new paradigms for how nations conceptualize, plan, and execute strategic competition.

The Fundamental Challenge: Bridging Human Wisdom and Machine Intelligence

The Core Tension

At the heart of the AI-driven strategic culture transformation lies a fundamental challenge that military leaders worldwide are grappling with: how to harness the unprecedented capabilities of artificial intelligence while preserving the essentially human elements of strategic thinking that have determined victory and defeat throughout history.

Traditional strategic culture has been built around distinctly human capabilities—intuition honed by experience, creative problem-solving in novel situations, and moral reasoning that considers broader implications beyond immediate tactical gains. These capabilities have evolved over centuries of military thought, from Sun Tzu's emphasis on knowing oneself and one's enemy to Clausewitz's recognition of war's essentially political nature. Strategic cultures have incorporated these human elements into institutional structures, decision-making processes, and operational concepts that reflect each nation's unique historical experience and values.

However, AI systems operate according to entirely different principles. They excel at pattern recognition, data processing, and optimization within defined parameters, but they struggle with the ambiguity, creativity, and moral reasoning that characterize human strategic thinking. As researchers at the Center for Security and Emerging Technology note, "AI systems have the ability to predict enemy behavior, anticipate vulnerabilities, weather and environmental conditions, assess mission strategies, and suggest alleviation plans" at speeds far beyond human capacity, yet they require "humans to make the ultimate decision" because machines cannot fully account for ethical principles, national security interests, and situational nuances that elude algorithmic analysis.

This tension manifests in multiple dimensions that are reshaping strategic cultures across the globe. Military leaders must now navigate the integration of AI capabilities that can accelerate decision-making from weeks to seconds while ensuring that the deliberative quality and moral foundation of strategic thinking are preserved. The challenge is not simply technical but cultural—requiring fundamental changes in how military institutions conceptualize the relationship between human judgment and machine intelligence.

The Expansion of AI Capabilities

The scope of AI integration in military strategic planning has expanded dramatically over the past few years, moving far beyond narrow applications to encompass comprehensive decision support systems that can influence strategy at the highest levels. Recent developments demonstrate just how thoroughly AI is penetrating strategic planning processes.

The Pentagon's Thunderforge project, awarded to Scale AI in March 2025, represents the most ambitious attempt yet to integrate AI into strategic planning. As described by Defense Innovation Unit program lead Bryce Goodman, "Today's military planning processes rely on decades-old technology and methodologies, creating a fundamental mismatch between the speed of modern warfare and our ability to respond." Thunderforge aims to bridge this gap by providing "AI-powered analysis and automation to operational and strategic planning, allowing decision-makers to operate at the pace required for emerging conflicts."

The system's capabilities extend across multiple domains of strategic planning. According to the Defense Innovation Unit, Thunderforge will support "mission-critical planning activities, including campaign development, theater-wide resource allocation, and strategic assessment." The platform leverages advanced large language models, AI-driven simulations, and interactive agent-based wargaming to enhance how the U.S. military prepares for and executes operations. Significantly, the system will initially be deployed with U.S. Indo-Pacific Command and U.S. European Command—the two combatant commands responsible for deterring America's most capable adversaries.

Scale AI CEO Alexandr Wang has described Thunderforge as "the flagship program within the DoD for AI-based military planning and operations," indicating the central role AI is expected to play in future strategic competition. The system represents what military planners call "agentic AI"—artificial intelligence capable of working through complex tasks autonomously to achieve assigned objectives, rather than simply responding to individual prompts.

This expansion of AI capabilities is creating what military strategists term "decision advantage"—the ability to process information and generate options faster than adversaries can adapt. As noted in recent military analyses, "At the operational level, we're seeing the rise of autonomous systems and swarm warfare, where decision cycles are measured in seconds, not days." Thunderforge is specifically "designed to bridge this gap by accelerating strategic planning without sacrificing human oversight."

Challenges of Human-Machine Integration

The integration of AI into strategic planning is revealing complex challenges that go far beyond technical implementation. Military leaders and researchers are discovering that successful human-AI collaboration requires fundamental changes in how strategic cultures approach decision-making, risk assessment, and organizational learning.

One of the most significant challenges is what researchers term "automation bias"—the tendency for human decision-makers to over-rely on AI recommendations without sufficient critical evaluation. As military AI experts note, "If decision-makers lean too heavily on AI without developing their own critical thinking and problem-solving skills, they may struggle to adapt when AI fails or is unreliable due to limitations in contingency planning or gaps in its training data."

This challenge is particularly acute in strategic planning, where the stakes of poor decisions can be catastrophic. The Center for Security and Emerging Technology identifies several specific risks in military AI applications: "False expectations with LLMs. Large language models (LLMs) are powerful tools, but they must be applied with care as they can mislead users by confidently presenting incorrect information, fabricating justifications, or increasing user acceptance of erroneous recommendations."

The challenge of maintaining human agency in strategic decision-making becomes even more complex when considering the speed advantages that AI systems provide. Military planners note that "one touted military advantage of AI is the increase in tempo of decision-making it would give a user over their adversary." However, this increased tempo often creates additional risks, particularly for civilian populations. "Increased tempo often creates additional risks to civilians, which is why techniques that reduce the tempo, such as 'tactical patience', are employed to reduce civilian casualties."

Another critical challenge is ensuring that human strategic leaders maintain their cognitive capabilities even as they become more dependent on AI support. Military education experts emphasize that "In high-stakes environments with unpredictable variables, leaders must exercise sound judgment without solely depending on algorithmic outputs. True leadership goes beyond accessing information—it requires the ability to analyze, question, and integrate knowledge to make decisive and effective choices under pressure."

The integration challenge is further complicated by the need to maintain interoperability across allied forces. NATO's revised AI strategy from July 2024 specifically identifies "increasing interoperability between AI systems throughout the Alliance" as a key priority, recognizing that future coalition operations will require AI systems from different nations to work together seamlessly. However, as military analysts note, this creates new complications: "Computer vision systems from different nations potentially contradicting each other and slowing down targeting cycle rather than speeding it up" if not properly coordinated.

New Strategic Concepts Emerging

Intelligentized Warfare: The Chinese Vision

Perhaps no country has been more systematic in developing new strategic concepts around AI integration than China, which has articulated a comprehensive vision of what it terms "intelligentized warfare" (�����). This concept represents a fundamental evolution in Chinese strategic thinking and provides insight into how AI is reshaping strategic cultures globally.

Chinese military strategists view intelligentized warfare as the next stage in military evolution, following mechanization and informatization. According to Chinese President Xi Jinping's directive to the People's Liberation Army, the goal is to "accelerate the integrated development of mechanization, informatization, and intelligentization" by 2027. This represents more than technological upgrades—it embodies a new strategic culture that places AI at the center of military effectiveness.

The PLA's conception of intelligentized warfare emphasizes information dominance over kinetic destruction. As Chinese military analysts explain, "AI should be used kinetically and non-kinetically to dominate information systems and networks, to effectively paralyze an opponent's joint force."

This approach reflects a strategic culture that prioritizes cognitive and informational effects over traditional measures of military power.

Central to the Chinese vision is the concept of "cognitive advantage"—the ability to adapt one's system-of-systems faster than one's adversary. As analyzed by military experts, "The PLA's objective is to use AI algorithms, machine learning, human-machine teaming, and autonomous systems collaboratively to paralyze its adversaries. The ultimate goal for the Chinese military appears to be cognitive advantage."

This focus on cognitive effects has led to the development of sophisticated information warfare capabilities. Recent analysis reveals that "China's cognitive warfare doctrine extends beyond military applications, leveraging AI-enhanced psychological operations to influence domestic and foreign audiences alike." The PLA has reportedly deployed "sophisticated machine learning models to create personalized propaganda" and views "AI-powered psychological warfare as a key enabler for securing strategic objectives."

The Chinese approach to intelligentized warfare also emphasizes swarming tactics and system destruction. Military analysts note that the PLA is developing capabilities for "large amounts of low-cost unmanned systems to saturate enemies" and pursuing "system destruction warfare targeting vulnerable links between sensors and platforms." This represents a strategic culture that seeks to achieve effects through coordinated AI-enabled systems rather than individual platforms.

However, the Chinese approach also reveals potential vulnerabilities. Analysts note the risk of "political distortions of military assessments due to authoritarian reporting culture" and suggest that "Beijing's overreliance on AI, political distortions of military assessments, and systemic resistance to delivering bad news to Chinese Communist Party General Secretary Xi Jinping could lead to strategic missteps in a future conflict."

Multi-Domain Precision Warfare

The integration of AI is enabling new operational concepts that transcend traditional domain boundaries. Military strategists are developing approaches to "multi-domain precision warfare" that leverage AI to integrate operations across land, sea, air, space, and cyber domains simultaneously.

This concept represents a fundamental shift from platform-centric to network-centric warfare. As military analysts explain, AI enables "multi-domain precision warfare" that aims to "integrate AI and big data analysis with precision strikes to identify and target enemy weaknesses." The approach requires generating, transmitting, and processing vast amounts of information to identify vulnerabilities in enemy operating systems and networks—capabilities that are only possible with AI integration.

The U.S. military's Combined Joint All-Domain Command and Control (CJADC2) initiative exemplifies this transformation. The program seeks to create an AI-powered global network that can coordinate forces across all domains in real-time. As described by defense contractors working on the project, the goal is to create systems that provide "decision dominance and overmatch" by enabling commanders to "make better and faster decisions even with imperfect information."

Advanced edge computing capabilities are crucial to this vision. Military planners are deploying AI systems that can operate at the "tactical edge"—forward-deployed locations where traditional connectivity to centralized data centers may be limited or compromised. As one expert explains, "Edge computing enables AI-enhanced drones or autonomous vehicles to capture and process data during complex missions without delays or central processing. This level of autonomy limits the risk to human lives while preserving real-time access to intelligent information."

The F-35 fighter aircraft provides an early example of multi-domain AI integration. The aircraft's Multifunction Advanced Data Link (MADL) system "uses high bandwidth to connect and share environmental data between each participating F-35s" and "automatically links their systems and their shared sensor data to create a uniform analysis of any threats, targets, or unforeseen change to the airspace—more complete than any one aircraft could gather on its own."

Edge AI and Distributed Operations

The concept of "edge AI" is revolutionizing how military forces approach distributed operations. Unlike traditional centralized command structures that rely on connectivity to distant data centers, edge AI enables autonomous decision-making at forward locations where communications may be intermittent or compromised.

Edge AI addresses one of the most significant vulnerabilities in modern military operations: dependence on network connectivity. As military technologists explain, "In remote or hostile environments, network connectivity can be intermittent or unreliable. Tactical edge cloud systems are designed to work in disconnected or low-bandwidth conditions, allowing them to continue functioning even when the network is not available."

This capability is particularly crucial for operations against sophisticated adversaries who may target communications networks. Military analysts note that edge AI systems are designed to be "EW-proof due to AI's capability to process electronic threats and operate independently" without requiring constant connectivity to centralized systems.

The U.S. Army's "Digital Soldier" initiative exemplifies this approach. The program uses "small, connected sensors that travel with soldiers to enable local data sharing and processing across network participants" and provides "individuals and squadrons with rapid information to ensure overmatch and safety." As military leaders explain, the networking effect multiplies capabilities exponentially—"the effect of networking is the n-squared—and this effect continues to increase as the number of network participants go up."

Edge AI also enables new forms of autonomous operations. Military systems equipped with edge AI can "process data to make split-second decisions" and "navigate complex terrain, identify threats, and even communicate with each other in real-time" without requiring human oversight for routine functions. This capability is essential for operations in contested environments where traditional command and control systems may be compromised.

Cognitive Warfare and Information Dominance

The emergence of AI-enabled cognitive warfare represents perhaps the most significant evolution in strategic thinking since the development of nuclear weapons. This concept moves beyond traditional kinetic effects to target human decision-making and perception directly.

NATO's recent analysis defines cognitive warfare as efforts to "exploit human cognition and technology to disrupt, undermine, influence, or modify human decision-making." The concept has gained prominence as military strategists recognize that "effective cognitive warfare requires seamless synchronization and coordination to manipulate human cognition, influencing decision-makers' comprehension of the information environment and their resolve to engage in conflict."

Chinese military theorists have been particularly innovative in developing cognitive warfare concepts. They view it as an extension of the "Three Warfares" doctrine (psychological warfare, public opinion warfare, and legal warfare), enhanced through AI automation. Recent analysis reveals that "by integrating AI into these domains, the PLA is enhancing its ability to automate disinformation campaigns, conduct large-scale social media manipulation, and employ deepfake technology to distort reality."

The targeting of cognitive processes represents a fundamental shift in strategic thinking. As military analysts note, Chinese strategists believe that "if the PLA gains an advantage in the development of AI technology, the PLA will be able to control human cognition, which is the lifeline of national security." This approach seeks to achieve strategic objectives by shaping adversary perceptions and decision-making rather than through traditional military defeat.

AI enables cognitive warfare operations at unprecedented scale and sophistication. Military researchers describe scenarios where "ultra-small intelligent unmanned systems, perhaps simulating a small animal, can enter the rooms of the highest decision-makers (president, members of Congress, combatant commanders) without being detected" to influence decision-making through information manipulation.

The implications of cognitive warfare extend beyond military operations to threaten democratic institutions themselves. NATO analysts warn that "AI-enabled disinformation, information operations and gender-based violence" could "affect the outcome of elections, sow division and confusion across the Alliance, demobilize and demoralize societies and militaries in times of conflict as well as lower trust in institutions and authorities of importance to the Alliance."

Cultural Adaptation Challenges

Speed vs. Deliberation: The Tempo Dilemma

One of the most profound challenges facing strategic cultures adapting to AI integration is managing the tension between AI-enabled speed and the deliberative processes that have traditionally characterized sound strategic decision-making. This tension strikes at the heart of strategic culture itself, challenging fundamental assumptions about how nations should approach consequential decisions.

The speed advantages offered by AI systems are genuinely revolutionary. Military analysts note that modern AI-enabled operations feature "decision cycles measured in seconds, not days," a compression of timeline that would have been inconceivable in previous eras of warfare. The Pentagon's Thunderforge system exemplifies this transformation, promising to accelerate strategic planning processes that traditionally took weeks or months into timeframes measured in minutes or hours.

However, this acceleration creates profound risks for the quality of strategic decision-making. Military ethicists and strategic theorists warn that increased tempo "often creates additional risks to civilians" and can lead to strategic miscalculations when human judgment is compressed beyond reasonable limits. The International Committee of the Red Cross emphasizes that slower decision-making processes traditionally "allow both the system and the user the extra time to develop more options" and reduce the likelihood of errors that could have catastrophic consequences.

Historical precedents underscore the importance of deliberative strategic decision-making. The most frequently cited example involves Soviet Lieutenant Colonel Stanislav Petrov, who in 1983 prevented a potential nuclear war by choosing to question an automated warning system that had detected what appeared to be incoming U.S. missiles. Petrov's decision to delay and investigate rather than follow protocol and authorize an immediate counterattack prevented what could have been global catastrophe. This incident highlights the crucial role that human deliberation and skepticism play in strategic decision-making, particularly in high-stakes situations.

The challenge for contemporary strategic cultures is developing approaches that harness AI's speed advantages while preserving the deliberative quality that prevents catastrophic errors. Military strategists are experimenting with various models of human-AI integration that attempt to balance these competing demands. Some propose "tactical patience" approaches that use AI to rapidly generate and analyze options while preserving human decision-making authority over final choices. Others advocate for "human-in-the-loop" systems that require human authorization for critical decisions even when AI systems recommend immediate action.

The resolution of this tension is crucial because it will determine whether AI integration enhances or degrades strategic decision-making quality. Military analysts emphasize that "the true test of future leadership lies in striking the right balance: leveraging AI's capabilities while preserving the independent thought, adaptability, and critical reasoning essential for command in today's volatile geopolitical environment."

Prediction vs. Adaptability: The Innovation Paradox

AI systems excel at pattern recognition and prediction based on historical data, but this strength creates a potential vulnerability in strategic cultures that become overly dependent on predictive analytics. The challenge is maintaining strategic adaptability and creative problem-solving capabilities even as AI systems provide increasingly sophisticated forecasts of future scenarios.

Military AI systems are designed to identify patterns in vast datasets and extrapolate future trends based on historical precedents. As military technologists explain, AI can "predict enemy behavior, anticipate vulnerabilities, weather and environmental conditions, assess mission strategies, and suggest alleviation plans" with remarkable accuracy when operating within established parameters. These capabilities offer genuine advantages in routine operational planning and resource allocation.

However, the most significant strategic challenges often involve novel situations that fall outside historical patterns. Military historians note that decisive strategic innovations—from blitzkrieg tactics to asymmetric warfare approaches—succeed precisely because they confound enemy predictions based on past experience. Over-reliance on AI prediction systems could potentially reduce strategic cultures' capacity for the kind of creative adaptation that has historically determined victory in major conflicts.

This challenge is particularly acute when facing adversaries who understand and can exploit AI prediction systems. Military analysts warn about "adversarial attacks" where enemies deliberately manipulate the data environment to deceive AI systems. "Deceptive attacks have become increasingly common, as illustrated by cases involving image classification algorithms that are deceived into perceiving images that aren't there, confusing the meaning of images, and mistaking a turtle for a rifle, for instance." In strategic contexts, such deception could lead AI-dependent forces to make catastrophic miscalculations.

The solution requires strategic cultures that use AI predictions as tools for enhancing human decision-making rather than replacing human strategic creativity. Military educators emphasize that successful leaders must be prepared to "outthink, not just out-tech, the competition" and maintain the ability to "command both machine intelligence and the power of the human mind." This requires deliberate efforts to preserve human cognitive capabilities even as AI systems become more sophisticated.

Some military organizations are experimenting with "red team" exercises specifically designed to test the adaptability of human-AI teams when confronting novel scenarios that fall outside AI training parameters. These exercises reveal both the potential and limitations of AI-enhanced strategic planning, helping military cultures develop more nuanced approaches to human-machine collaboration.

Centralization vs. Decentralization: The Command Paradox

AI integration is creating complex tensions in military command structures, simultaneously enabling both greater centralization through enhanced information processing and greater decentralization through autonomous edge systems. Strategic cultures must navigate these opposing pressures while maintaining effective command and control.

On one hand, AI systems enable unprecedented centralization of strategic planning and control. Systems like Thunderforge can process theater-wide data and generate comprehensive operational plans that account for resources, capabilities, and constraints across vast geographic areas. This capability allows senior commanders to maintain detailed awareness and control over operations that would have been impossible to coordinate centrally in previous eras.

Advanced AI systems can "consolidate, analyze, and act on vast quantities of theater-wide data across the warfighting functions" in real-time, providing senior leaders with comprehensive situational awareness that extends from strategic to tactical levels. This enhanced visibility enables more precise coordination and resource allocation across large-scale operations.

However, the same AI technologies that enable centralization also make decentralized autonomous operations more effective. Edge AI systems can operate independently of centralized command structures, making local decisions based on immediate tactical situations without requiring authorization from higher headquarters. Military analysts note that these systems can "navigate complex terrain, identify threats, and even communicate with each other in real-time" without human oversight.

This decentralization capability is particularly valuable in contested environments where communications with centralized command structures may be compromised. Autonomous systems equipped with edge AI can continue executing missions even when traditional command and control networks are disrupted. As military strategists explain, "tanks, drones, and even soldiers can be equipped with edge devices that process data to make split-second decisions" without requiring connectivity to distant command centers.

The challenge for strategic cultures is developing command philosophies that effectively leverage both centralized coordination and decentralized autonomy. This requires new approaches to mission command that provide clear intent and boundaries for autonomous systems while preserving human authority over strategic objectives and rules of engagement.

Some military organizations are experimenting with "mission command evolution as dialogue between man and machine," where AI systems provide recommendations and autonomous capabilities within parameters established by human commanders. This approach seeks to combine the speed and precision of autonomous systems with the judgment and accountability of human leadership.

The resolution of this tension will likely determine the effectiveness of future military operations, as successful strategic cultures will need to optimize the benefits of both centralized coordination and decentralized autonomy while maintaining human control over strategic objectives and ethical constraints.

Alliance Implications and Interoperability Challenges

NATO's AI Integration Struggles

The integration of AI into NATO operations presents unique challenges that highlight the complex intersection of technology and alliance politics. Unlike national military organizations that can implement unified AI strategies, NATO must coordinate AI development and deployment across 32 member nations with varying technological capabilities, strategic cultures, and political systems.

NATO's revised AI strategy, released in July 2024, acknowledges these challenges while attempting to create a framework for alliance-wide AI integration. The strategy identifies "increasing interoperability between AI systems throughout the Alliance" as a central priority, recognizing that future coalition operations will require seamless integration of AI systems developed by different nations with different technical standards and operational concepts.

The interoperability challenge is more complex than traditional military compatibility issues because AI systems often operate in ways that are opaque even to their operators. Military analysts warn that "computer vision systems from different nations potentially contradicting each other and slowing down targeting cycle rather than speeding it up" if AI systems from different countries interpret the same data differently or operate according to conflicting algorithms.

NATO's approach emphasizes the development of common standards and testing protocols through its Defense Innovation Accelerator for the North Atlantic (DIANA) network. The alliance is working to establish "Alliance-wide AI Testing, Evaluation, Verification & Validation landscape" that can ensure AI systems from different member nations can work together effectively. However, this coordination effort faces significant challenges in terms of both technical complexity and political sensitivity.

The technical challenges are compounded by the rapid pace of AI development in the commercial sector. NATO's strategy recognizes that "innovation ecosystems implicating different actors and faster technology lifecycles than typically included in traditional capability development systems" require new approaches to technology adoption and integration. The alliance must balance the need for common standards with the desire to leverage rapidly evolving commercial AI capabilities.

Political challenges add another layer of complexity. Different NATO members have varying levels of AI development capabilities and different approaches to AI governance and ethics. The strategy acknowledges that "varying levels of ethical and threat assessments as well as the differing infrastructure capabilities of member states" complicate efforts to establish common standards.

Moreover, "the alliance has no enforcement mechanism for AI standards," making coordination largely dependent on voluntary compliance.

Recent tensions over technology sharing highlight these political dimensions. Military analysts note that "22 NATO members excluded from U.S. 'trusted countries' list affecting technology sharing" creates complications for alliance AI integration. Some allies have protested these restrictions, with "Poland and Israel protesting exclusion from trusted country lists" and arguing that such limitations undermine alliance cohesion.

Information Sharing and Classification Challenges

The integration of AI into alliance operations creates unprecedented challenges for information sharing and classification protocols. Traditional classification systems were designed for static documents and clearly defined intelligence products, but AI systems continuously generate new analyses and insights that challenge existing security frameworks.

Military security experts identify "classification by compilation" as a particular concern with AI systems. As one analyst explains, AI can "piece together classified information from unclassified sources," creating situations where ostensibly unclassified AI outputs contain sensitive information that was derived from classified inputs. This challenge is compounded by the fact that "no one's come up with great answers for appropriate classification of AI products."

The problem is further complicated by the speed at which AI systems operate. Traditional security review processes assume human analysts will have time to review and classify intelligence products before they are shared. However, AI systems can generate new analyses "constantly creating new analyses" that outpace security review capabilities. This creates a fundamental tension between the speed advantages that AI provides and the security protocols necessary to protect sensitive information.

These classification challenges have direct implications for alliance operations. NATO partners need to share AI-generated intelligence and coordinate AI-enabled operations, but existing classification systems may prevent effective collaboration. The alliance must develop new approaches to classification that can accommodate the dynamic nature of AI-generated intelligence while maintaining appropriate security protections.

Some military organizations are experimenting with automated classification systems that use AI to identify sensitive information in AI-generated products. However, these systems face their own challenges in terms of accuracy and reliability. False positives could unnecessarily restrict information sharing, while false negatives could compromise operational security.

The resolution of these challenges is crucial for effective alliance AI integration. Without reliable mechanisms for sharing AI-generated intelligence and coordinating AI-enabled operations, the potential benefits of AI integration may not be realized in coalition contexts where interoperability is essential.

Competitive Dynamics Within Alliances

The development of AI capabilities is creating new competitive dynamics within alliances that challenge traditional concepts of burden-sharing and collective security. Nations with advanced AI capabilities may gain significant advantages in intelligence collection, operational planning, and battlefield effectiveness, potentially creating imbalances within alliance structures.

These competitive dynamics are particularly visible in the relationship between the United States and its allies. U.S. restrictions on AI technology sharing, including export controls on advanced semiconductors and AI software, have created tensions with alliance partners who feel excluded from critical technological developments. Recent policy decisions treating "disparate countries such as India and Tuvalu the same" in terms of technology access have frustrated allies who expected preferential treatment based on their alliance relationships.

Smaller NATO allies face particular challenges in maintaining relevance within an AI-enabled alliance structure. Countries like Canada are developing AI strategies specifically designed to "remain credible and interoperable" with larger allies while leveraging their limited resources effectively. The challenge is contributing meaningfully to collective security without attempting to match the AI capabilities of larger powers.

Some allies are responding by pursuing specialization strategies, focusing on specific AI domains where they can develop competitive advantages rather than attempting comprehensive AI development programs. This approach seeks to create complementary rather than redundant capabilities, with different allies leading in different AI applications such as cyber operations, autonomous systems, or intelligence analysis.

However, specialization also creates dependencies that may affect alliance dynamics. If certain allies become the primary providers of specific AI capabilities, they may gain disproportionate influence over alliance decision-making in those domains. This could undermine the principle of sovereign equality that has traditionally characterized NATO operations.

The NATO Innovation Fund, established as "the world's first multi-sovereign venture capital fund" with €1 billion in funding, represents an attempt to address these competitive dynamics by creating shared investment in AI development. However, the fund's impact on reducing competitive tensions within the alliance remains to be seen.

The resolution of these competitive dynamics will likely determine the long-term cohesion of AI-enabled alliances. Successful alliances will need to develop mechanisms that allow member nations to contribute according to their capabilities while maintaining the shared values and mutual interests that bind alliance relationships together.

Contemporary Examples and Implementation

U.S. Military AI Integration

The U.S. military's approach to AI integration provides perhaps the most comprehensive example of how traditional strategic cultures are adapting to artificial intelligence. With an AI budget request of $1.8 billion for fiscal 2025 and over 180 identified military AI applications, the Pentagon's efforts illustrate both the potential and challenges of large-scale AI adoption.

The Thunderforge project represents the most ambitious AI integration effort to date. Awarded to Scale AI in March 2025, the project will initially deploy to U.S. Indo-Pacific Command and U.S. European Command—the two combatant commands responsible for deterring China and Russia respectively. The system's deployment to these critical theaters demonstrates the Pentagon's confidence in AI's potential to provide strategic advantages against peer competitors.

Thunderforge's capabilities extend across the full spectrum of strategic planning. According to Defense Innovation Unit Director Doug Beck, the system provides "the foundation for the next generation of AI-driven military decision-making, enhancing the Joint Force's ability to plan, adapt, and respond to emerging challenges at machine speed." The platform leverages advanced large language models, AI-driven simulations, and interactive agent-based wargaming to accelerate processes that traditionally took weeks or months.

However, U.S. implementation efforts also reveal significant challenges in adapting traditional military culture to AI capabilities. Military leaders emphasize that human oversight remains essential, with safeguards ensuring that "if a weapon is involved or a maneuver risks human life, a human decision-maker is always in the loop." This reflects an attempt to preserve human accountability while leveraging AI efficiency.

The Pentagon has also implemented AI in more specialized applications across multiple domains. Intelligence analysis has been revolutionized through systems that can process vast amounts of data from satellite imagery, social media, and sensor networks. As military analysts explain, work that previously required "many individuals, over dozens or hundreds of hours, can be done in milliseconds to seconds by AI systems."

Phase Two of the Pentagon's military AI push focuses specifically on generative AI applications. Marines are now using "generative AI to scour intelligence through chatbot interface similar to ChatGPT," representing a significant shift from traditional intelligence analysis methods. However, this implementation also highlights classification challenges, as these systems must operate within security constraints that traditional AI applications were not designed to accommodate.

European Approaches to AI Military Integration

European military organizations are pursuing AI integration strategies that reflect both technological capabilities and distinct strategic cultures emphasizing democratic oversight and ethical frameworks. The European Union's AI Act, which takes a "human-centric, risk-based model," contrasts with more "flexible, innovation-friendly standards" pursued by the United States, reflecting different approaches to balancing innovation with regulation.

France has been particularly active in developing military AI capabilities while maintaining emphasis on human control and ethical considerations. French President Emmanuel Macron has emphasized that "the future of AI is a political stake, of sovereignty and strategic dependence," reflecting European concerns about technological sovereignty in the face of U.S. and Chinese AI dominance.

The European approach emphasizes "digital sovereignty" and "jurisdictional independence" over competitive advantage, seeking to develop AI capabilities that reduce dependence on non-European technologies while maintaining compatibility with alliance partners. This approach reflects strategic cultures that prioritize regulatory leadership and ethical frameworks as forms of technological influence.

Multiple European Defence Fund and PESCO (Permanent Structured Cooperation) projects are integrating AI into military capabilities, but with explicit emphasis on maintaining human oversight and democratic accountability. These projects reflect European strategic cultures that view technology as serving political objectives rather than determining them.

Germany's approach to military AI exemplifies European emphasis on ethical frameworks. German military AI development explicitly incorporates international humanitarian law considerations and emphasizes explainability and human oversight in AI system design. This approach reflects a strategic culture that views technological capability as inseparable from legal and ethical compliance.

The United Kingdom has pursued a middle path between American innovation-focused approaches and European regulation-focused approaches. UK defense officials emphasize that "human political control maintained at all times" regardless of AI use, reflecting British strategic culture's emphasis on political accountability for military actions.

Chinese Military AI Implementation

China's approach to military AI implementation reflects its authoritarian strategic culture and long-term planning capabilities. The People's Liberation Army's systematic development of "intelligentized warfare" capabilities demonstrates how different political systems can enable different approaches to AI integration.

Chinese military AI development benefits from what analysts term "authoritarian advantages" in technology adoption. The Chinese system's ability to make "rapid AI integration" decisions without extensive democratic consultation allows faster implementation of AI capabilities across military organizations. This approach reflects a strategic culture that prioritizes technological advancement over deliberative processes.

The PLA's Military-Civil Fusion strategy enables systematic integration of commercial AI advances into military capabilities. Recent analysis reveals that "China now globally dominates 57 of 64 critical technologies" related to AI development, providing a broad technological base for military applications. This dominance reflects sustained investment and strategic planning that leverages both state and commercial AI development.

Chinese military AI applications emphasize information dominance and cognitive effects over kinetic capabilities. The PLA has developed sophisticated capabilities for "AI-enhanced psychological operations to influence domestic and foreign audiences" and deployed "sophisticated machine learning models to create personalized propaganda." These applications reflect a strategic culture that prioritizes information warfare and political effects.

However, Chinese AI implementation also reveals potential vulnerabilities. Analysts note risks of "political distortions of military assessments due to authoritarian reporting culture" and suggest that over-reliance on AI systems without adequate real-world testing could lead to strategic miscalculations. The Chinese military's lack of recent combat experience compounds these risks.

The PLA's approach to human-AI integration differs significantly from Western approaches. Chinese military writings contain limited discussion of "keeping humans in the loop," and some analysts suggest the PLA "could prove less averse to the prospect of taking humans out of the loop" compared to Western militaries. This reflects a strategic culture with different assumptions about human agency and technological authority.

Recent developments include the establishment of the PLA Information Support Force (formerly Strategic Support Force) to coordinate AI and information warfare capabilities across military domains. This organizational change reflects Chinese strategic culture's emphasis on centralized coordination and systematic integration of emerging technologies.

Key Insight Development

The transformation of strategic cultures through AI integration represents a fundamental shift in how nations conceptualize and conduct strategic competition. This transformation goes far beyond the adoption of new technologies—it requires conscious cultural evolution that preserves human wisdom while harnessing machine intelligence.

The evidence suggests that successful adaptation requires what might be termed "algorithmic strategic cultures"—institutional approaches that seamlessly integrate human strategic thinking with AI capabilities while maintaining human agency over fundamental strategic decisions. Nations that develop such cultures will likely gain significant advantages in future strategic competition, while those that fail to adapt risk strategic irrelevance.

However, the development of algorithmic strategic cultures is not simply a matter of technological implementation. It requires fundamental changes in military education, institutional structures, and decision-making processes that reflect new realities of human-machine collaboration. Most importantly, it requires conscious choices about how to preserve the essentially human elements of strategic thinking—creativity, ethical reasoning, and adaptability—that will continue to determine success in strategic competition.

The nations and alliances that navigate this transformation most successfully will be those that view AI integration as an opportunity to enhance rather than replace human strategic capabilities. This requires strategic cultures capable of continuous learning and adaptation, rather than rigid adherence to traditional approaches. The stakes of this transformation could not be higher, as the next generation of strategic cultures will determine the balance of power and the nature of strategic competition for decades to come.

Verifiable References and Sources

1. Belfer Center for Science and International Affairs. "AI's New Frontier in War Planning: How AI Agents Can Revolutionize Military Decision-Making." October 11, 2024. https://www.belfercenter.org/research-analysis/ais-new-frontier-war-planning-how-ai-agents-can-revolutionize-military-decision
2. Breaking Defense. "AI for war plans: Pentagon innovation shop taps Scale AI to build 'Thunderforge' prototype." March 5, 2025. http://breakingdefense.com/2025/03/ai-for-war-plans-pentagon-innovation-shop-taps-scale-ai-to-build-thunderforge-prototype/
3. Center for Security and Emerging Technology. "AI for Military Decision-Making: Harnessing the Advantages and Avoiding the Risks." April 1, 2025. https://cset.georgetown.edu/publication/ai-for-military-decision-making/
4. Center for Strategic and International Studies. "Calibrating NATO's Vision of AI-Enabled Decision Support." October 11, 2024. https://www.csis.org/analysis/calibrating-natos-vision-ai-enabled-decision-support
5. Chinese Military Innovation in Artificial Intelligence. Center for a New American Security. June 17, 2019. https://www.cnas.org/publications/congressional-testimony/chinese-military-innovation-in-artificial-intelligence
6. Defense Innovation Unit. "DIU's Thunderforge Project to Integrate Commercial AI-Powered Decision-Making." https://www.diu.mil/latest/dius-thunderforge-project-to-integrate-commercial-ai-powered-decision-making
7. Decrypt. "US Military Taps 'Thunderforge' AI for Wargaming and Planning Operations." March 5, 2025. https://decrypt.co/308829/us-military-taps-thunderforge-ai-for-wargaming-and-planning-operations
8. Foreign Policy Research Institute. "AI Dependence and Political Blind Spots Undermine Beijing's War Strategy." March 18, 2025. https://www.fpri.org/article/2025/03/ai-dependence-and-political-blind-spots-undermine-beijings-war-strategy/
9. Frontiers in Big Data. "Cognitive warfare: a conceptual analysis of the NATO ACT cognitive warfare exploratory concept." November 1, 2024. https://www.frontiersin.org/

journals/big-data/articles/10.3389/fdata.2024.1452129/full
10. International Committee of the Red Cross. "Artificial intelligence in military decision-making: supporting humans, not replacing them." August 29, 2024. https://blogs.icrc.org/law-and-policy/2024/08/29/artificial-intelligence-in-military-decision-making-supporting-humans-not-replacing-them/
11. International Committee of the Red Cross. "Algorithms of war: The use of artificial intelligence in decision making in armed conflict." February 7, 2025. https://blogs.icrc.org/law-and-policy/2023/10/24/algorithms-of-war-use-of-artificial-intelligence-decision-making-armed-conflict/
12. Modern War Institute. "Warfare at the Speed of Thought: Balancing AI and Critical Thinking for the Military Leaders of Tomorrow." February 21, 2025. https://mwi.westpoint.edu/warfare-at-the-speed-of-thought-balancing-ai-and-critical-thinking-for-the-military-leaders-of-tomorrow/
13. NATO. "NATO releases revised AI strategy." July 10, 2024. https://www.nato.int/cps/en/natohq/news_227234.htm
14. NATO. "Summary of NATO's revised Artificial Intelligence (AI) strategy." July 10, 2024. https://www.nato.int/cps/en/natohq/official_texts_227237.htm
15. NATO Parliamentary Assembly. "2024 - NATO AND AI - REPORT." https://www.nato-pa.int/document/2024-nato-and-ai-report-clement-058-stc
16. Recorded Future. "Artificial Eyes: Generative AI and China's Military Intelligence." January 2025. https://www.recordedfuture.com/research/artificial-eyes-generative-ai-chinas-military-intelligence
17. Scale AI. "Introducing Thunderforge: AI for American Defense." https://scale.com/blog/thunderforge-ai-for-american-defense
18. Sentient Digital. "Military Applications of AI in 2024." June 20, 2024. https://sdi.ai/blog/the-most-useful-military-applications-of-ai/
19. The Register. "Pentagon to give AI agents a role in planning, operations." March 5, 2025. https://www.theregister.com/2025/03/05/dod_taps_scale_to_bring/
20. The Washington Post. "Pentagon contract 'Thunderforge' will make AI tools for military planning." March 5, 2025. https://www.washingtonpost.com/technology/2025/03/05/pentagon-ai-military-scale/
21. Total Military Insight. "The Role of AI in Military Planning for Strategic Decisions." July 16, 2024. https://totalmilitaryinsight.com/ai-in-military-planning/
22. War on the Rocks. "Chinese Debates on the Military Utility of Artificial Intelligence." June 5, 2020. https://warontherocks.com/2020/06/chinese-debates-on-the-military-utility-of-artificial-intelligence/

23. War on the Rocks. "New Tech, New Concepts: China's Plans for AI and Cognitive Warfare." April 15, 2022. https://warontherocks.com/2022/04/new-tech-new-concepts-chinas-plans-for-ai-and-cognitive-warfare/

Chapter 4: America's Democratic Dilemma: Innovation Versus Control

Balancing Open Innovation with Strategic Competition

Opening Scene: The Silicon Valley-Pentagon Convergence

In March 2025, at the Defense Innovation Unit's Silicon Valley headquarters in Mountain View, California, a decision was made that would epitomize America's fundamental challenge in the age of artificial intelligence. Defense officials announced that Scale AI, a San Francisco-based company founded by the son of two Los Alamos National Laboratory physicists, had been awarded the prime contract for "Thunderforge"—the Pentagon's flagship program to integrate AI agents directly into military planning and operations.

The scene was striking in its contradictions. In a conference room overlooking the tech campuses of Google, Meta, and Apple, military officials in uniform sat alongside venture capitalists in casual dress discussing how to deploy AI systems that could process "vast amounts of data" to generate battlefield recommendations at "machine speed." Scale AI CEO Alexandr Wang, just 27 years old, spoke confidently about how his company would "transform today's military operating process and modernize American defense," while Defense Innovation Unit Director Doug Beck emphasized the urgent need for "machine speed" decision-making to match the pace of modern warfare.

The $200-million Thunderforge contract, developed alongside partners including Microsoft and defense startup Anduril, represents more than just another government procurement. It embodies the central tension at the heart of American strategic culture in the AI age: how does a democratic society harness the revolutionary potential of artificial intelligence for national defense while maintaining the oversight, accountability, and human control that define democratic governance?

Unlike authoritarian systems where AI integration can proceed through centralized decree, the United States must navigate a complex ecosystem of competing interests, ethical concerns, and democratic institutions. The Pentagon must secure cutting-edge AI capabilities from Silicon Valley companies whose employees have historically protested military applications of their technology. Congress must provide oversight of AI systems that operate at speeds beyond traditional legislative review. And military leaders must balance the promise of AI-enabled "decision advantage" against the risk of losing human agency in life-and-death decisions.

As DIU's Bryce Goodman noted in announcing Thunderforge, there exists a "fundamental mismatch between the speed of modern warfare and our ability to respond." Current military planning processes depend on "decades-old technology and methodologies" while adversaries deploy AI systems that compress decision cycles from days to seconds. The democratic challenge is acute:

How can the United States adapt its strategic culture to harness AI's speed and precision while preserving the deliberation and accountability that democratic governance requires?

Strategic Culture Evolution Analysis

From "Overwhelming Force" to "Decision Advantage"

For generations, American strategic culture has been built around the concept of overwhelming technological and material superiority. From the World War II arsenal of democracy through the precision-guided munitions revolution of the 1990s, the United States military has consistently sought to compensate for potential numerical disadvantages through technological dominance. This approach worked when technology cycles moved slowly enough for democratic institutions to understand, debate, and oversee new capabilities.

AI fundamentally disrupts this model. The Pentagon's Data and AI Adoption Strategy, released in 2023, explicitly shifts the focus from overwhelming force to "decision advantage"—the ability to observe, orient, decide, and act faster than adversaries. As Deputy Defense Secretary Kathleen Hicks emphasized in announcing the fiscal 2025 budget, the Pentagon is investing $1.8 billion in AI to "equip department leaders with data to make better decisions faster."

This represents a profound evolution in American strategic thinking. The traditional American way of war—mobilizing massive resources over extended timeframes—gives way to a new paradigm where victory depends on algorithmic speed and cognitive superiority. The Pentagon's fiscal 2025 budget request of $1.8 billion for AI, while representing flat growth from fiscal 2024 due to congressional spending caps, reflects this strategic priority even under fiscal constraints. According to Pentagon budget documents, this funding "supports efforts to deliver and adopt responsible AI/ML-enabled capabilities on secure and reliable platforms, workforce development, and DOD-wide data management and modernization efforts."

The transformation extends beyond budget allocations to fundamental concepts of military effectiveness. Where previous generations of military leaders thought in terms of platforms and firepower, today's emerging strategic culture increasingly values data fusion, algorithmic analysis, and human-machine teaming. The Pentagon now oversees more than 685 AI-related projects spanning from battlefield assessments to predictive maintenance, each representing a shift from human-centered to AI-augmented decision-making.

This evolution creates new tensions within American strategic culture. The traditional emphasis on human judgment and deliberative planning confronts AI systems that can process intelligence and generate recommendations in seconds. As one defense official noted, "the planning and operational process for the U.S. military has not evolved since Napoleon," highlighting the massive cultural adaptation required to integrate AI capabilities effectively.

Democratic Oversight Challenge

The Pentagon's rapid AI adoption poses unprecedented challenges to democratic accountability. According to Government Accountability Office assessments, as of 2024 the Defense Department maintains over 180 documented instances of AI military applications, ranging from intelligence analysis to autonomous systems to strategic planning support. This represents a dramatic increase from the 685 unclassified AI projects tracked by GAO in 2021, illustrating the accelerating pace of AI integration across military operations.

The challenge of congressional oversight becomes apparent when examining the scope and complexity of these systems. Take the example of Marines using "generative AI to scour intelligence through chatbot interface similar to ChatGPT"—a capability that exemplifies how AI is being embedded throughout military operations in ways that traditional oversight mechanisms struggle to address. Congress faces the challenge of providing meaningful oversight of AI systems that generate new analyses continuously, often at classification levels that limit transparency.

The classification problem is particularly acute. As noted in Pentagon assessments, AI systems are "constantly creating new analyses" at a pace that outstrips traditional security review processes. This creates what officials describe as "classification by compilation"—where AI can piece together classified insights from unclassified data sources, fundamentally challenging existing information security frameworks. As one Pentagon official noted, "no one's come up with great answers for appropriate classification of AI products."

Representative Sara Jacobs, a member of the House Armed Services Committee, highlighted the democratic oversight dilemma in congressional hearings: "Humans are fallible, and at least we have accountability processes in place. But it's hard to have the same kind of accountability if a decision is made purely by an algorithm." This tension between AI speed and democratic deliberation appears throughout congressional engagement with military AI.

The GAO has repeatedly emphasized this accountability gap. In a March 2024 assessment, GAO Director for Defense Capabilities and Management Alissa Czyz noted that "as of November 2023, DOD had not fully implemented" any of the recommendations from GAO's 2022 review of Pentagon AI adoption. The oversight challenges include incomplete information about the number of AI activities and investments in the Defense Department's AI portfolio, unclear roles and responsibilities among military services for AI collaboration, and insufficient frameworks for evaluating AI system performance and accountability.

Congress has attempted to address these challenges through legislation like the Federal AI Governance and Transparency Act, introduced by Representatives James Comer and Jamie Raskin in 2024. The bill would require agencies to notify individuals when government decisions are meaningfully influenced by AI and provide options for human review through existing appeals processes. However, implementing such requirements for military AI systems operating at classified levels and machine speeds presents significant practical challenges.

Current AI Initiatives and Budget

The Pentagon's approach to AI integration in fiscal 2025 reflects both ambition and constraint. The $1.8 billion AI budget request, while flat from fiscal 2024 due to the Fiscal Responsibility Act's spending caps, represents a strategic commitment to AI adoption even under fiscal pressure. According to defense officials, this maintains the trajectory from $874 million in fiscal 2022 to $1.1 billion in fiscal 2023, demonstrating consistent prioritization of AI capabilities.

The budget supports several major initiatives that illustrate the breadth of American AI military integration. The Thunderforge program, led by Scale AI, represents the Pentagon's first major foray into integrating AI agents directly into operational military planning. Initially deploying to U.S. Indo-Pacific Command and U.S. European Command, Thunderforge will provide "AI-assisted planning capabilities, decision support tools, and automated workflows" for theater-level military operations.

Beyond Thunderforge, the Pentagon's AI investments span multiple domains. In June 2025, the Army's Enterprise Large Language Model Workspace, developed by Ask Sage, grew from zero users to 19,000 in just 45 days, leading to a $10 million contract expansion to combatant commands and Pentagon offices. This rapid adoption illustrates both the demand for AI tools and the speed at which AI capabilities are being integrated throughout military operations.

The Department's Chief Digital and Artificial Intelligence Office (CDAO), established in 2022 through consolidation of four separate AI organizations, has overseen dramatic growth in AI contracting. According to Brookings Institution analysis, Defense Department AI-related contracts increased from $190 million in 2022 to $557 million in 2023, with the total potential value of AI federal contracts rising from $355 million to $4.6 billion in the same period.

Recent contract awards illustrate the expanding scope of military AI applications. In July 2025, the Pentagon awarded contracts worth up to $200 million each to four major AI companies: Google, OpenAI, Anthropic, and Elon Musk's xAI. These contracts, managed by the Chief Digital and AI Office, aim to accelerate adoption of "advanced AI capabilities to address critical national security challenges" across warfighting and enterprise domains.

The Pentagon has also significantly expanded its investment in Project Maven, raising the contract ceiling for Palantir Technologies' Maven Smart System to nearly $1.3 billion through 2029. This represents a major expansion of the AI-powered intelligence analysis program that began controversially in 2017 and has since become central to military intelligence operations.

However, budget analysis reveals the challenge of scaling AI capabilities within fiscal constraints. According to Obviant analysis of Pentagon budget data, the Defense Department allocated approximately $25.2 billion for programs incorporating AI and autonomous systems in fiscal 2025—about 3% of the total $850 billion defense budget. The majority ($21.8 billion) supports research, development, testing, and evaluation, while only $3.4 billion is designated for procurement, highlighting the early-stage nature of many AI military applications.

Strategic Culture Tensions Deep Dive

Innovation vs. Security

The fundamental tension between innovation and security permeates every aspect of America's AI strategic culture. The Pentagon's relationship with Silicon Valley epitomizes this challenge: the Defense Department needs cutting-edge AI capabilities that can only be developed by private companies operating in competitive markets, yet integrating these commercial technologies into military systems requires security measures that can stifle innovation.

This tension became visible in the evolution of tech company policies toward military work. For years after 2018, when Google employees successfully protested the company's involvement in Project Maven (AI analysis of drone footage), major tech companies maintained policies restricting military applications of their AI systems. OpenAI explicitly prohibited use of its models for "weapons development" or "military and warfare" in its usage policies through early 2024.

The shift began in January 2024 when OpenAI quietly softened these restrictions, changing its policy to forbid using AI technology to "harm yourself or others" rather than blanket military restrictions. By March 2025, OpenAI was announcing the Thunderforge partnership, marking what MIT Technology Review called "OpenAI's military pivot." This transformation reflects the broader Silicon Valley evolution from opposition to military applications toward active embrace of defense contracts.

The economic drivers behind this shift are significant. Training and operating large language models costs hundreds of millions of dollars, and consumer revenue alone cannot sustain these investments. The Pentagon, with its $850 billion annual budget and increasing AI focus, represents an attractive market for companies seeking returns on massive AI investments. As one AI executive told Fortune magazine, the defense market represents "basically building the airplane while they're flying it, so it's a massive land grab."

However, this commercial motivation creates security challenges. The Pentagon must balance access to cutting-edge commercial AI capabilities against the risk of dependence on companies whose primary loyalty is to shareholders rather than national security. The challenge is particularly acute given that the same large language models used for military applications are also available to commercial customers, including potential adversaries.

The classification and export control challenges compound these security concerns. Traditional export controls focus on hardware and software exports, but AI capabilities are increasingly embedded in cloud services and accessible through APIs. The Pentagon's new rules forcing cloud providers to keep "50% of compute power inside US" while capping international installations represent one attempt to address these challenges, but they also risk driving international customers toward non-American providers.

Speed vs. Deliberation

The collision between AI speed and democratic deliberation represents perhaps the most fundamental challenge to American strategic culture. As Pentagon officials note, AI systems can compress decision cycles from days to seconds, potentially transforming military operations but challenging traditional decision-making processes that emphasize deliberation and oversight.

The Thunderforge system exemplifies this tension. Scale AI's technology will provide military commanders with AI-generated course-of-action recommendations based on analysis of "vast amounts of data" including "latest locations of potential targets, unit readiness reports, transport routes and timetables, the load-carrying capacities of road bridges and seaports, stockpiles of everything from ammunition to spare parts." The system promises to accelerate strategic planning by automating data analysis that currently requires "hundreds of hours" of human staff work.

But this speed comes with risks. As defense analysts note, the pressure for AI-enabled rapid response could undermine the deliberative quality that has traditionally characterized American strategic decision-making. The concern is particularly acute given automation bias—the tendency for humans to defer to machine recommendations, particularly under time pressure.

Military leaders recognize this challenge. The Army's approach to generative AI emphasizes maintaining human oversight while leveraging AI speed. Army CIO Leonel Garciga described the rapid adoption of the Army's AI workspace—from zero to 19,000 users in 45 days—while emphasizing that these tools augment rather than replace human decision-making.

The nuclear domain illustrates the ultimate stakes of this tension. While the Pentagon maintains official policy requiring "humans in the loop for nuclear weapons employment," the pressure to match adversary decision speeds creates constant tension. As noted in congressional hearings, "an ICBM takes at most half an hour to reach its target," creating time pressure that could favor AI-assisted decision-making even in nuclear scenarios.

Congressional efforts to address this tension include the "Block Nuclear Launch by Autonomous AI Act," introduced by Senator Ed Markey and Representative Ted Lieu in 2023. The bipartisan legislation would prohibit federal funding for any nuclear weapon launch by automated systems without "meaningful human control," codifying existing Pentagon policy while preventing future policy changes that might remove human oversight under pressure.

Human Control vs. Machine Efficiency

The challenge of maintaining human control while leveraging machine efficiency pervades every aspect of AI military integration. Pentagon policy officially maintains "human in the loop" requirements for critical decisions, but the practical implementation of this principle proves complex as AI systems become more sophisticated and integrated throughout military operations.

The evolution of "human in the loop" concepts illustrates this complexity. Early automation required humans to initiate every action, but modern AI systems can operate with "human on the loop" oversight—where humans monitor but don't directly control every decision—or even "human out of the loop" for certain automated functions. The Pentagon's approach varies by domain and risk level, creating a spectrum of human-AI integration rather than a simple binary choice.

In practice, this creates challenging scenarios. When Marines use AI chatbots to analyze intelligence, human operators make the final analytical judgments but rely heavily on AI processing of vast data sets. When AI systems recommend target priorities or deployment options, human commanders maintain decision authority but operate under time pressures that can bias them toward accepting machine recommendations.

The challenge is particularly acute in defensive systems. Pentagon policy explicitly states that autonomous weapons restrictions "do not apply to cyber" operations, recognizing that cyber defense often requires automated responses at machine speeds. Similarly, existing systems like the Navy's Close-In Weapons System (CIWS) already operate autonomously to defend ships against incoming missiles, illustrating that some level of automation has long been accepted for defensive purposes.

Former Pentagon Chief Digital and AI Officer Craig Martell acknowledged this evolution, noting that "it was no longer possible to keep a human in the loop for every autonomous decision" given the speed and scale of modern military operations. This pragmatic recognition drives ongoing efforts to develop what military planners call "meaningful human control"—ensuring human accountability and oversight while allowing necessary automation.

The Defense Innovation Unit's emphasis on making AI reasoning "legible" reflects one approach to this challenge. By ensuring that AI systems can "trace their logic, assess confidence levels," human operators can maintain meaningful oversight even when operating at machine speeds. However, as AI systems become more sophisticated, even this traceability becomes challenging to implement in practice.

Key Challenges Section

Public-Private AI Cooperation

The relationship between Silicon Valley and the Pentagon has fundamentally transformed since the 2018 Google employee protests over Project Maven. What began as widespread tech industry resistance to military applications has evolved into active competition for lucrative defense contracts, creating new challenges for managing public-private AI cooperation.

The transformation is visible in recent contract awards. In July 2025, the Pentagon simultaneously awarded $200 million contracts to Google, OpenAI, Anthropic, and xAI—companies that just years earlier maintained policies against military AI applications. This represents what industry observers call a "military AI gold rush" driven by both national security imperatives and commercial pressures.

The challenge lies in managing these relationships while maintaining appropriate boundaries. Defense contracts provide tech companies with stable revenue streams to support expensive AI development, but they also create potential conflicts between commercial and national security interests. When the same AI models serve both civilian customers and military applications, questions arise about data security, technology transfer, and long-term reliability of commercial partners.

Employee resistance within tech companies adds another layer of complexity. Even as company executives embrace defense contracts, many employees continue to oppose military applications of their work. OpenAI employees raised ethical concerns on internal forums following the company's December 2024 partnership with defense contractor Anduril, asking how the company could ensure its technology wouldn't be used for offensive purposes beyond defensive anti-drone systems.

Google has faced similar internal tensions. Despite the company's return to defense work, employee protests continue over contracts like Project Nimbus with the Israeli government. In 2024, Google fired over 50 employees following protests against military-related contracts, illustrating the ongoing friction between corporate strategy and employee values.

The Pentagon must navigate these dynamics while securing reliable access to cutting-edge AI capabilities. Defense Innovation Unit Director Doug Beck acknowledged the challenge, noting that the Pentagon needs to work with commercial providers while ensuring "accuracy and reliability" and maintaining "human oversight" over AI systems. The solution involves careful contract structuring, security requirements, and ongoing oversight to balance commercial innovation with national security needs.

The talent retention issue compounds these challenges. AI expertise is scarce and highly valued in commercial markets, making it difficult for the Pentagon to maintain in-house capabilities sufficient for meaningful oversight of commercial AI systems. This creates what experts call the "AI oversight gap"—where government lacks sufficient technical expertise to effectively evaluate and manage commercial AI capabilities being integrated into military systems.

Alliance Interoperability

America's AI strategic culture must account for alliance relationships that require sharing capabilities while maintaining competitive advantages. The challenge is particularly acute given that AI systems from different nations may "contradict each other and slow down targeting cycles rather than speed them up," as NATO assessments have noted.

The Pentagon's AI initiatives increasingly focus on ensuring interoperability with allied systems. The Thunderforge system, for example, will initially deploy to U.S. Indo-Pacific Command and U.S. European Command—both of which operate extensively with allied forces. Ensuring that American AI systems can share data and coordinate with allied technologies while maintaining security and operational advantages requires careful technical and policy coordination.

NATO's revised AI strategy, released in July 2024, acknowledges these challenges by calling for an "Alliance-wide AI Testing, Evaluation, Verification & Validation landscape" through the DIANA network. The €1 billion NATO Innovation Fund represents an attempt to coordinate allied AI development while maintaining interoperability. However, implementing common standards across different national AI development approaches proves challenging in practice.

The classification and technology sharing challenges are significant. U.S. restrictions treat "disparate countries such as India and Tuvalu the same" in technology access controls, according to congressional assessments, creating friction with key allies. Poland and Israel have protested exclusion from U.S. "trusted countries" lists for AI technology sharing, illustrating the tension between security requirements and alliance cooperation.

The human-in-the-loop coordination challenge is particularly complex. While the United States maintains official policies requiring human control for nuclear weapons and other critical systems, ensuring that allied AI systems operate under compatible constraints requires extensive coordination. As one NATO assessment noted, autonomous systems must be "programmed to operate together without relying on common sense or relationships between human operators."

The timeline mismatch between national AI development and alliance coordination creates additional challenges. While commercial AI development proceeds at venture capital speeds measured in quarters, alliance coordination operates on diplomatic timelines measured in years. The Pentagon must balance the imperative for rapid AI adoption against the need for allied coordination and consensus-building.

Ethical Constraints

The tension between military effectiveness and democratic values pervades American AI strategic culture. While authoritarian systems can potentially integrate AI without extensive ethical constraints, the United States must balance AI capabilities against democratic principles, human rights commitments, and international legal obligations.

The Pentagon's official AI ethical principles, adopted in 2020 and reaffirmed in subsequent policy documents, emphasize five key principles: responsible, equitable, traceable, reliable, and governable AI. These principles require that AI systems be designed with human oversight, operate without harmful biases, provide explainable decisions, function reliably under operational conditions, and remain under appropriate human control.

Implementing these principles in practice proves challenging. The requirement for "traceable" AI conflicts with the "black box" nature of some machine learning systems that provide accurate results through opaque processes. The requirement for "equitable" AI faces challenges when AI systems must make rapid targeting or threat assessment decisions based on limited data that may contain historical biases.

The congressional response includes multiple legislative efforts to codify ethical constraints. The "Block Nuclear Launch by Autonomous AI Act" represents the most dramatic example, legally prohibiting AI from making nuclear launch decisions. The Federal AI Governance and Transparency Act would require notification when AI systems influence government decisions affecting individuals, though implementing such requirements for classified military systems presents practical challenges.

International coordination on AI ethics creates additional constraints and opportunities. The United States has joined over 40 nations in endorsing a political declaration on responsible military use of AI, establishing principles for ethical AI development and deployment. However, these commitments are non-binding and must compete with operational imperatives and adversary AI development that may not operate under similar ethical constraints.

The civilian harm mitigation challenge illustrates the complexity of implementing ethical constraints in practice. AI systems used for targeting and collateral damage assessment must operate under strict legal and ethical guidelines, but they must also function effectively in combat environments where perfect information is unavailable and time constraints are severe. The White House's National Security Memorandum on AI, signed in October 2024, addresses this by requiring "continuous rigorous testing" of AI systems used for collateral damage assessment and maintaining "human oversight by trained personnel."

Future Trajectory Analysis

Maintaining Technological Superiority

America's strategy for maintaining AI technological superiority must navigate the inherent tensions between democratic governance and competitive AI development. Unlike authoritarian systems that can direct resources and suppress opposition through centralized control, the United States must maintain technological leadership through a complex ecosystem of commercial innovation, democratic oversight, and alliance cooperation.

The challenge is evident in current competitive dynamics. While the Pentagon invests $1.8 billion annually in AI, China's reported AI military investments significantly exceed American levels when accounting for civil-military fusion and state-directed development. The Pentagon's approach seeks to compensate for this potential resource disadvantage by leveraging America's competitive commercial AI ecosystem, but this strategy requires maintaining an attractive business environment for AI companies while imposing necessary security constraints.

The Department's future strategy increasingly emphasizes creating a competitive AI ecosystem rather than relying on single providers. The simultaneous $200 million contract awards to Google, OpenAI, Anthropic, and xAI in July 2025 illustrate this approach—creating competition among AI providers while ensuring the Pentagon isn't dependent on any single commercial partner. This diversification strategy aims to maintain innovation incentives while reducing technological dependency risks.

The talent and expertise challenge will likely intensify. AI capabilities depend on human expertise that is globally mobile and highly valued in commercial markets. The Pentagon must compete with tech companies offering significantly higher compensation while providing meaningful work that attracts top AI talent to national security applications. Programs like the Army's Technical Fellows initiative and partnerships with universities represent attempts to build sustainable AI expertise within government.

The timeline compression challenge requires fundamental changes to traditional defense acquisition processes. While commercial AI development operates on six-month innovation cycles, traditional defense procurement operates on multi-year timelines. The Pentagon's increasing use of prototyping contracts, like the Thunderforge system, represents an attempt to accelerate AI adoption while maintaining appropriate oversight and testing.

Preserving Democratic Oversight

The future of democratic oversight of AI military capabilities will require new institutions and processes designed for AI's speed and complexity. Traditional congressional oversight mechanisms, designed for platform-based military systems with predictable capabilities, must adapt to AI systems that continuously learn and evolve in ways that can surprise even their designers.

The classification challenge will likely require new approaches to transparency and accountability. Current security review processes cannot keep pace with AI systems that "constantly create new analyses" at machine speeds. Future solutions may involve automated classification systems, new categories of information sharing with congressional oversight bodies, or restructured security clearance processes that allow more rapid legislative review of AI capabilities.

The expertise gap between congressional oversight bodies and AI technical reality will require sustained investment in legislative branch technical capabilities. The Government Accountability Office's investment in AI capabilities for its own operations represents one model—developing internal expertise that can effectively evaluate executive branch AI implementations. The Congressional Budget Office and Congressional Research Service face similar needs for AI expertise to support legislative oversight.

The real-time oversight challenge may require new institutional arrangements. Traditional congressional hearings and budget reviews operate on annual cycles, but AI systems operate continuously. Future oversight may require standing technical advisory bodies, real-time monitoring systems, or automated reporting mechanisms that can provide ongoing insight into AI system performance and impacts without compromising operational security.

The transparency vs. security tension will likely drive development of new classification and disclosure frameworks specifically designed for AI systems. Current classification systems, designed for hardware and traditional software, struggle with AI systems that derive capabilities from training data, algorithmic approaches, and operational experience. Future frameworks may need to distinguish between core AI architectures (which may require classification) and performance metrics or operational impacts (which may support greater transparency).

Alliance Cohesion

America's future AI strategy must balance national technological advantages with alliance coordination requirements. As AI becomes increasingly central to military operations, the risk grows that technological asymmetries could undermine alliance effectiveness even as they provide national advantages.

The interoperability imperative will drive increasing standardization of AI interfaces and protocols among allied forces. NATO's AI testing and validation framework represents an early attempt to ensure compatible AI systems, but future requirements will likely demand much deeper technical coordination. This may require sharing some proprietary AI capabilities with allies or developing common AI architectures that sacrifice some competitive advantages for operational compatibility.

The technology sharing framework will need fundamental revision to account for AI characteristics. Traditional export controls focus on preventing adversary access to specific technologies, but AI capabilities often derive from training data, algorithmic innovations, and operational experience that are difficult to control through traditional export mechanisms. Future frameworks may need to focus on AI service access and capability sharing rather than technology transfer in traditional senses.

The burden-sharing evolution will likely see allied specialization in different aspects of AI military capabilities. Rather than each ally attempting to develop comprehensive AI capabilities, future arrangements may involve division of labor where different allies lead in different AI domains—intelligence processing, logistics optimization, autonomous systems, or cyber operations. This specialization could enhance overall alliance capabilities while making each member more dependent on others for complete AI capabilities.

The timeline synchronization challenge will require new mechanisms for coordinating AI development across different national acquisition systems. American commercial AI development, European regulatory-focused approaches, and Pacific ally technology strategies operate on different timelines and priorities. Future coordination may require new international institutions or treaty frameworks specifically designed for AI technology cooperation and standard-setting.

Innovation Ecosystem Sustainability

The long-term sustainability of America's AI innovation ecosystem requires balancing competitive commercial markets with national security requirements. The current model of leveraging commercial AI development for military applications faces potential sustainability challenges as AI capabilities mature and market dynamics evolve.

The commercial viability question will become increasingly important as AI development costs continue rising. Current AI leaders like OpenAI, Google, and Anthropic can sustain massive development investments through venture capital and corporate support, but long-term commercial viability requires sustainable revenue models. If commercial AI markets consolidate or prove unsustainable, the Pentagon may need to support AI development more directly, potentially requiring significant budget increases.

The technological dependency risk grows as AI systems become more sophisticated and specialized. If American AI capabilities become dependent on small numbers of commercial providers, national security vulnerabilities could emerge from business failures, foreign acquisitions, or other commercial disruptions. The Pentagon's multi-vendor strategy aims to mitigate these risks, but the high costs of advanced AI development may naturally drive market consolidation.

The international competition dimension will intensify as other nations develop competitive AI capabilities. China's state-directed AI development model, European regulatory approaches, and emerging AI capabilities in countries like India and Israel create both competitive pressures and potential partnership opportunities. America's future strategy must account for a multipolar AI landscape rather than assuming continued American dominance.

The workforce and education foundation requires sustained investment in AI-capable personnel across government, industry, and academia. The Pentagon's emphasis on ensuring "combat leaders understand AI basics" represents one aspect of this challenge, but the broader requirement includes developing AI expertise throughout government, maintaining educational institutions that produce AI-capable personnel, and creating career pathways that attract top talent to national security applications.

The regulatory and oversight framework must evolve to support rather than constrain beneficial AI innovation while preventing harmful applications. Future success requires regulatory approaches that can distinguish between beneficial and harmful AI uses, provide appropriate oversight without stifling innovation, and maintain democratic accountability while allowing necessary operational secrecy. This balance will require continued experimentation and adaptation as AI capabilities and applications continue evolving.

Contemporary Case Studies

OpenAI's Military Pivot and Changing Policies

OpenAI's transformation from AI safety advocate to military contractor illustrates the broader evolution of Silicon Valley's relationship with national security. Founded in 2015 with the mission to ensure that artificial intelligence "benefits humanity as a whole," OpenAI maintained explicit prohibitions against military use of its technologies through early 2024.

The company's original usage policies explicitly prohibited anyone from using its models for "weapons development" or "military and warfare." These restrictions reflected broader Silicon Valley skepticism about military applications following the 2018 Google employee protests over Project Maven. OpenAI CEO Sam Altman reinforced this position in public statements, suggesting there were applications he would "never do with the Department of Defense."

The shift began subtly in January 2024 when The Intercept reported that OpenAI had softened its usage restrictions. Rather than blanket prohibitions on military use, the new policies forbade using AI technology to "harm yourself or others" by developing weapons, injuring people, or destroying property. This change opened the door for defensive military applications while maintaining restrictions on offensive weapons development.

By October 2024, OpenAI published a blog post arguing that AI could "help protect people, deter adversaries, and even prevent future conflict" when deployed in appropriate national security contexts. The company explicitly argued that helping democratic countries maintain AI advantages would serve its mission of ensuring AI benefits humanity broadly.

The December 2024 partnership with defense contractor Anduril marked OpenAI's full entry into military applications. The partnership focuses on counter-drone systems, which OpenAI characterizes as defensive applications aligned with protecting American lives. However, internal employee messages viewed by The Washington Post revealed significant concern among OpenAI staff about the implications of partnering with a weapons manufacturer.

Employees questioned how OpenAI could ensure Anduril systems wouldn't be directed against human-piloted aircraft or deployed in other offensive ways. One employee noted the irony that Skynet, the fictional AI system from Terminator films, was originally designed for defensive purposes similar to current counter-drone applications. Others worried about reputational damage and the precedent of working with weapons manufacturers.

The January 2025 announcement of a direct $200 million Pentagon contract through the "OpenAI for Government" program completed the company's transformation. The contract, awarded by the Pentagon's Chief Digital and AI Office, provides the Defense Department with access to OpenAI's latest models for various national security applications. OpenAI CEO Sam Altman characterized the work as supporting those who "risk their lives to keep our families and our country safe."

This evolution reflects broader changes in Silicon Valley's approach to national security work. Economic pressures from massive AI development costs, combined with growing recognition of strategic competition with China, have overcome earlier employee resistance to military applications. The success of defense-focused companies like Palantir and Anduril in securing lucrative government contracts has demonstrated the commercial viability of defense AI applications.

However, the employee resistance illustrates ongoing tensions within tech companies about military applications. While executive leadership has embraced defense opportunities, significant portions of technical workforces remain uncomfortable with military applications of their work. This tension may affect talent retention and company culture as AI capabilities become increasingly central to both commercial and military applications.

Google's Evolution from Cautious to Military-Engaged AI Development

Google's relationship with military AI applications has undergone dramatic reversals that illustrate the complexity of managing public-private AI cooperation in democratic systems. The company's experience with Project Maven in 2018 established the template for Silicon Valley resistance to military applications, but recent contract awards demonstrate how commercial and strategic pressures have overcome earlier ethical objections.

Project Maven, launched in 2017, used Google's computer vision capabilities to analyze drone surveillance footage for the Pentagon. The program aimed to automate the labor-intensive process of identifying objects and people in drone video, potentially accelerating intelligence analysis and target identification. While Google insisted the technology was used only for "non-offensive purposes" such as analyzing footage rather than selecting targets, employee opposition was immediate and intense.

Nearly 5,000 Google employees signed petitions demanding cancellation of the Maven contract, arguing that the company's technology should not be used for military applications that could lead to harm. Dozens of employees resigned in protest, and internal employee activism forced CEO Sundar Pichai to cancel a planned vacation to address staff concerns. The pressure was so intense that Google ultimately declined to renew the Maven contract when it expired.

The controversy led Google to adopt formal "AI Principles" in 2018 that explicitly restricted military applications. The principles stated that Google would "not pursue" technologies "whose principal purpose or implementation is to cause or directly facilitate injury to people" or applications for "surveillance violating internationally accepted norms." The company explicitly committed to avoiding AI development for weapons systems.

However, the principles contained significant ambiguities that allowed for later reinterpretation. The restriction on technologies whose "principal purpose" is causing harm left room for dual-use applications where military capability was a secondary rather than primary purpose. The surveillance restrictions were limited to applications "violating internationally accepted norms" rather than all military surveillance applications.

These ambiguities became important as Google faced increasing competitive pressure from other tech companies willing to work with the Pentagon. While Google maintained restrictions on direct weapons development, competitors like Microsoft, Amazon, and Oracle secured lucrative cloud computing contracts worth billions of dollars. The commercial pressure intensified as AI development costs continued rising and defense spending on AI capabilities accelerated.

The transformation became visible in 2025 when Google secured a $200 million contract with the Pentagon's Chief Digital and AI Office alongside OpenAI, Anthropic, and xAI. The contract, announced in July 2025, represents Google's return to direct defense work after seven years of restrictions. The company characterizes its AI military work as supporting "non-offensive" applications and emphasizing defensive capabilities.

Google's current approach attempts to balance commercial opportunities with employee concerns by focusing on applications that can be characterized as protective or defensive rather than offensive. The company continues to face internal resistance, with Google DeepMind researchers protesting contracts with the Israeli government over concerns about military applications of AI technology.

The Google experience illustrates several broader challenges in managing public-private AI cooperation. Employee activism remains a significant constraint on tech company military engagement, even as executive leadership embraces defense opportunities. The definitional challenges around "offensive" versus "defensive" AI applications create ongoing ambiguities that can be reinterpreted as commercial pressures change.

The competitive dynamics among tech companies create pressure for companies to match competitors' willingness to work with the Pentagon or risk losing market share to more cooperative firms. As one industry observer noted, the current environment represents "a buyer and seller market" where employees have less influence over corporate decisions than during the initial Maven protests.

Congressional Debates over AI Military Applications

Congressional engagement with military AI applications reveals the challenges democratic institutions face in providing effective oversight of rapidly evolving technologies. Recent legislative debates illustrate both the bipartisan recognition of AI's strategic importance and the difficulty of crafting appropriate oversight mechanisms for AI systems that operate at machine speeds.

The "Block Nuclear Launch by Autonomous AI Act," introduced by Senator Ed Markey and Representative Ted Lieu in April 2023, represents the most dramatic congressional effort to constrain military AI applications. The bipartisan legislation would prohibit federal funding for any nuclear weapon launch by automated systems without "meaningful human control." While the bill codifies existing Pentagon policy, it aims to prevent future policy changes that might remove human oversight under operational pressure.

The legislation gained bipartisan support from senators including Jeff Merkley, Bernie Sanders, and Elizabeth Warren, along with representatives including Don Beyer and Ken Buck. The broad support reflects congressional concern about maintaining human control over the most consequential military decisions, even as AI capabilities advance rapidly.

However, implementing the bill's "meaningful human control" requirement proves complex in practice. Congressional hearings have revealed disagreement among experts about what constitutes adequate human oversight of AI systems that can process information and generate recommendations in seconds. The time pressure of nuclear decision-making—with ICBMs capable of reaching targets in thirty minutes—creates scenarios where human deliberation may be impractical.

The Federal AI Governance and Transparency Act, introduced by Representatives James Comer and Jamie Raskin in March 2024, takes a broader approach to AI oversight. The bipartisan legislation would require agencies to notify individuals when government decisions are meaningfully influenced by AI and provide options for human review through existing appeals processes. The bill also requires agencies to publish governance charters for high-risk AI systems.

Representative Sara Jacobs of the House Armed Services Committee has focused on the accountability challenges AI creates for democratic governance. In committee hearings, she emphasized that "humans are fallible, and at least we have accountability processes in place" for human decisions, but questioned "how to have the same kind of accountability if a decision is made purely by an algorithm."

Jacobs has proposed several approaches to maintaining democratic oversight of military AI, including setting guardrails on training data used for AI systems, protecting servicemembers' personally identifiable information, and establishing clear rules about when AI autonomy is permitted on the battlefield. Her proposals illustrate the detailed technical and policy questions that effective AI oversight requires.

The budget oversight challenge has proven particularly complex. While Congress appropriates $1.8 billion annually for Pentagon AI programs, the distributed nature of AI implementation across hundreds of projects makes traditional oversight mechanisms difficult to apply. Government Accountability Office assessments have repeatedly noted that Congress lacks "complete information about the number of AI activities and investments that comprise DOD's AI portfolio."

GAO Director Alissa Czyz noted in 2024 testimony that "as of November 2023, DOD had not fully implemented" any of the recommendations from GAO's 2022 review of Pentagon AI adoption. The implementation challenges include unclear roles and responsibilities among military services, insufficient inventory processes for AI projects, and inadequate frameworks for evaluating AI system performance.

Congressional efforts to address these oversight challenges face the fundamental tension between AI speed and democratic deliberation. Traditional congressional oversight operates on annual budget cycles and periodic hearing schedules, while AI systems operate continuously and evolve rapidly. Future oversight mechanisms may require new institutional arrangements specifically designed for AI's temporal characteristics.

The classification challenges add another layer of complexity to congressional oversight. Many AI applications operate at classified levels that limit congressional access to technical details about system capabilities and performance. While congressional intelligence committees have appropriate clearances, the technical complexity of AI systems makes effective oversight challenging even with security access.

These legislative debates illustrate the broader challenge democratic societies face in governing AI applications that operate at speeds and complexities beyond traditional institutional capacities. The solutions require both new legal frameworks and new institutional capabilities designed specifically for AI oversight rather than adapting existing mechanisms designed for previous technologies.

Key Insight Development

America's approach to AI strategic culture transformation reflects the inherent tensions and advantages of democratic governance in the age of artificial intelligence. Unlike authoritarian systems that can integrate AI through centralized direction, the United States must harness AI capabilities while maintaining the oversight, accountability, and competitive innovation that define democratic societies.

The central challenge lies in preserving human agency and democratic accountability while leveraging AI's speed and analytical capabilities. The Pentagon's emphasis on "meaningful human control" and requirements for "humans in the loop" represent attempts to maintain democratic principles while adopting revolutionary capabilities. However, implementing these principles in practice requires ongoing adaptation as AI capabilities advance and operational requirements evolve.

The American model's strength lies in its competitive innovation ecosystem. The private sector's AI innovation, driven by market competition and venture capital investment, provides capabilities that government-directed development might not achieve. The recent pivot of companies like OpenAI and Google toward military applications demonstrates how commercial incentives can ultimately align with national security requirements, even overcoming initial employee resistance.

However, this market-based approach creates vulnerabilities around technological dependence, security requirements, and long-term sustainability. The Pentagon's strategy of diversifying AI providers and maintaining competitive dynamics represents an attempt to harness market benefits while mitigating dependence risks. The success of this approach will depend on maintaining an attractive business environment for AI companies while imposing necessary security constraints.

The democratic oversight challenge requires fundamental adaptation of traditional accountability mechanisms. Congressional oversight, GAO reviews, and public transparency requirements must evolve to address AI systems that operate at machine speeds and generate new capabilities continuously. The development of new institutional arrangements specifically designed for AI governance will be essential for maintaining democratic control over military AI applications.

The alliance dimension adds another layer of complexity to America's AI strategic culture. Maintaining technological advantages while ensuring interoperability with allied forces requires careful balance between competitive secrecy and cooperative sharing. The success of American AI strategy will ultimately depend on creating alliance frameworks that enhance collective capabilities without undermining individual competitive advantages.

The ethical constraints that democratic governance imposes on AI development represent both a limitation and a source of strength. While these constraints may slow certain AI applications compared to less constrained adversaries, they also provide legitimacy and sustainability that pure efficiency-based approaches may lack. The long-term success of AI military applications will likely depend on maintaining public confidence through ethical development and democratic oversight.

The future trajectory of American AI strategic culture will be determined by how successfully democratic institutions adapt to govern AI capabilities while preserving the competitive innovation that provides technological advantages. This adaptation requires conscious strategic culture design rather than accidental evolution—a process that must begin immediately while the window for shaping AI development remains open.

Sources and References

1. "AI for war plans: Pentagon innovation shop taps Scale AI to build 'Thunderforge' prototype," *Breaking Defense*, March 5, 2025. Available at: http://breakingdefense.com/2025/03/ai-for-war-plans-pentagon-innovation-shop-taps-scale-ai-to-build-thunderforge-prototype/
2. "Introducing Thunderforge: AI for American Defense," *Scale AI Blog*, 2025. Available at: https://scale.com/blog/thunderforge-ai-for-american-defense
3. "Pentagon contract 'Thunderforge' will make AI tools for military planning," *The Washington Post*, March 5, 2025. Available at: https://www.washingtonpost.com/technology/2025/03/05/pentagon-ai-military-scale/
4. "Why the Pentagon didn't request higher funding for AI in fiscal 2025," *DefenseScoop*, March 11, 2024. Available at: https://defensescoop.com/2024/03/11/pentagon-ai-budget-request-2025/
5. "Pentagon's priority on AI spending could shield it from cuts," *Roll Call*, March 4, 2024. Available at: https://rollcall.com/2024/03/05/pentagons-priority-on-ai-spending-could-shield-it-from-cuts/
6. "U.S. Military Spending on AI Surges," *TIME*, March 29, 2024. Available at: https://time.com/6961317/ai-artificial-intelligence-us-military-spending/
7. "Pentagon AI office, Army award Ask Sage $10M for GenAI expansion," *Breaking Defense*, June 20, 2025. Available at: https://breakingdefense.com/2025/06/pentagon-ai-office-army-award-ask-sage-10m-for-genai-expansion/
8. "Follow the Money: What the Pentagon's Budget Data Tells Us About AI and Autonomy Adoption," *Obviuant Analysis*, March 31, 2025. Available at: https://maggiegray.us/p/

follow-the-money-what-the-pentagons

9. "Anthropic, Google, OpenAI and xAI granted up to $200 million for AI work from Defense Department," *CNBC*, July 14, 2025. Available at: https://www.cnbc.com/2025/07/14/anthropic-google-openai-xai-granted-up-to-200-million-from-dod.html
10. "Pentagon boosts budget for Palantir's AI software in major expansion of Project Maven," *SpaceNews*, May 22, 2025. Available at: https://spacenews.com/pentagon-boosts-budget-for-palantirs-ai-software-in-major-expansion-of-project-maven/
11. "Markey, Lieu, Beyer, and Buck Introduce Bipartisan Legislation to Prevent AI From Launching a Nuclear Weapon," *U.S. Senator Ed Markey*, April 26, 2023. Available at: https://www.markey.senate.gov/news/press-releases/markey-lieu-beyer-and-buck-introduce-bipartisan-legislation-to-prevent-ai-from-launching-a-nuclear-weapon
12. "House Oversight leaders introduce government AI transparency bill," *Nextgov/FCW*, March 6, 2024. Available at: https://www.nextgov.com/artificial-intelligence/2024/03/house-oversight-leaders-introduce-government-ai-transparency-bill/394722/
13. "How new White House AI memo impacts, and restricts, the Pentagon," *Breaking Defense*, October 28, 2024. Available at: https://breakingdefense.com/2024/10/how-new-white-house-ai-memo-impacts-and-restricts-the-pentagon/
14. "OpenAI's new defense contract completes its military pivot," *MIT Technology Review*, January 24, 2025. Available at: https://www.technologyreview.com/2024/12/04/1107897/openais-new-defense-contract-completes-its-military-pivot/
15. "OpenAI employees question the ethics of military deal with startup Anduril," *The Washington Post*, December 7, 2024. Available at: https://www.washingtonpost.com/technology/2024/12/06/openai-anduril-employee-military-ai/
16. "Big Tech joins the military, with Meta, Palantir and OpenAI in front," *Quartz*, 2025. Available at: https://qz.com/tech-ai-military-pentagon-meta-google-openai
17. "Government watchdog prepares to assess Pentagon's revamped AI strategy," *DefenseScoop*, February 26, 2024. Available at: https://defensescoop.com/2023/12/21/gao-prepares-to-assess-pentagons-revamped-ai-strategy/
18. "Artificial Intelligence: GAO's Work to Leverage Technology and Ensure Responsible Use," *U.S. Government Accountability Office*, January 30, 2024. Available at: https://www.gao.gov/products/gao-24-107237
19. "Pentagon awards multiple companies $200M contracts for AI tools," *Nextgov/FCW*, July 14, 2025. Available at: https://www.nextgov.com/acquisition/2025/07/pentagon-awards-multiple-companies-200m-contracts-ai-tools/406698/
20. "The new 'land grab' for AI companies, from Meta to OpenAI, is military contracts," *Fortune*, November 27, 2024. Available at: https://fortune.com/2024/11/27/ai-

companies-meta-llama-openai-google-us-defense-military-contracts/

Chapter 5: China's Algorithmic Authoritarianism: AI as State Power Multiplier

From Mechanization to Intelligentization

Opening Scene: The Vision of Intelligentized Warfare

The scene unfolds in the Great Hall of the People in Beijing, October 2022. Chinese Communist Party General Secretary Xi Jinping steps to the podium at the 20th National Congress, his words carrying the weight of strategic transformation for the world's largest military. "We must speed up the development of unmanned, intelligent combat capabilities," Xi declares, his voice echoing through the vast chamber filled with Party officials and military leaders. The phrase "intelligent" (���) appears three times in his speech—a deliberate emphasis that signals China's commitment to what military strategists call "intelligentization," the integration of artificial intelligence into every facet of warfare.

This moment represents more than policy announcement; it marks the culmination of a strategic vision that Xi has been developing since taking power. Six months earlier, in March 2022, Xi had called on the People's Liberation Army (PLA) to "raise the presence of combat forces in new domains and of new qualities," setting the stage for a military transformation unlike any in modern history. The concept of intelligentization, which emerged officially in China's 2019 National Defense White Paper, now sits at the apex of the Party's military modernization strategy.

In the months following Xi's speech, PLA strategists would articulate their belief that intelligentization represents changes "as significant as those seen in WWII and the Gulf War, if not bigger." Unlike the mechanized warfare that defined the 20th century or the information warfare that emerged in the 1990s, intelligentized warfare aims to transcend the physical and information domains to operate in what Chinese military theorists call the "cognitive space"—where effective decision-making and complex thinking provide decisive advantage.

The implications ripple far beyond China's borders. As Xi speaks, PLA researchers are already testing AI systems that can process battlefield intelligence from multiple sources, create "dynamic kill networks" across domains, and execute strikes with minimal human intervention. The vision he articulates is not merely about upgrading weapons systems; it represents a fundamental reimagining of warfare itself, one where algorithms and human cognition merge to create new forms of strategic power.

This is China's bid to not merely compete with American military dominance but to fundamentally alter the nature of strategic competition itself. Through intelligentization, Beijing seeks to transform the PLA from a force that has historically struggled to match Western technological sophistication into one that could leapfrog ahead by harnessing the revolutionary potential of artificial intelligence. The question is not whether this transformation will reshape global strategic culture, but how quickly and how completely it will do so.

Strategic Vision Framework: The Three-Stage Revolution

The Path to Military Modernization

China's approach to military AI integration rests on a carefully structured three-stage modernization framework that Xi Jinping has elevated to the status of national strategy. This progression—from mechanization to informatization to intelligentization—represents more than technological evolution; it embodies China's strategic logic for achieving what Xi calls a "world-class" military by mid-century.

Mechanization forms the foundation, encompassing the acquisition of modern platforms and equipment capable of delivering firepower against adversaries. Chinese strategists recognize this as the "physical space" of warfare, drawing lessons from what they view as the apex of platform-centric warfare during World War II, when the United States dominated through mechanized formations like carrier battle groups. The PLA announced it had achieved basic mechanization in 2020, though significant portions of China's massive military remain reliant on older systems.

Informatization represents the second benchmark, focusing on linking military systems to networks such as GPS, advanced communications, and battlefield management systems. This phase emerged from China's profound shock at witnessing American technological dominance during the 1991 Gulf War, when the United States demonstrated the devastating potential of precision-guided weapons and networked forces. Chinese military observers concluded that informatization operates in the "information space" and is essential for collecting and efficiently processing battlefield information. The Gulf War became a watershed moment that determined China's subsequent decades of military modernization efforts.

Intelligentization, the culminating phase, integrates artificial intelligence, quantum computing, big data, and other emerging technologies into the joint force. Chinese strategists position this as operating in the "cognitive space," where decision-making speed and complex reasoning provide decisive advantages. According to the Center for a New American Security, "In 2020, China set a new goal to 'accelerate the integrated development of mechanization, informatization, and intelligentization' by 2027," meaning Beijing aims to progress on all three stages simultaneously rather than sequentially.

This parallel development strategy reflects both the urgency of China's strategic timeline and the pragmatic recognition that different parts of the PLA exist at different technological levels. As Hudson Institute analysts observe, this mirrors historical precedent: "In 1940, Germany defeated France in just 42 days by blitzkrieg, utilizing tanks. At that time, only a small percentage of the German army was mechanized, and the majority of the German army was still an old-fashioned force dependent on horses and foot soldiers." China's approach suggests that even partial intelligentization of select PLA units could provide revolutionary battlefield advantages.

Intelligentization as Revolutionary Change

Chinese military theorists frame intelligentization not as an incremental improvement but as a fundamental transformation in the character of warfare. The PLA's National Defense University and other premier military institutions have produced extensive research arguing that AI-enabled warfare will be qualitatively different from all previous forms of conflict.

The scope of this transformation is evident in China's official defense spending allocations. The CSIS analysis reveals that China's pursuit of defense technologies encompasses quantum imaging, navigation, and radar applications to enhance intelligence, surveillance, and reconnaissance capabilities, alongside position, navigation, and timing systems. The U.S. Defense Department's 2024 China Military Power Report notes that "The PLA is pursuing next-generation combat capabilities based on its vision of future conflict, which it calls 'intelligentized warfare,' defined by the expanded use of AI, quantum computing, big data and other advanced technologies at every level of warfare."

Central to this vision is the concept of "multi-domain precision warfare," which Chinese strategists see as their answer to American dominance in conventional military domains. Rather than attempting to match U.S. capabilities platform-for-platform, China's intelligentized warfare doctrine emphasizes "systems destruction warfare"—targeting the vulnerable information links between sensors and platforms rather than the platforms themselves. This approach reflects a fundamental shift from a focus on physical destruction to one emphasizing cognitive and informational effects.

The PLA anticipates what its strategists call a "singularity on the battlefield" where "human cognition can no longer keep pace with the speed of decision-making." This represents more than technological acceleration; it envisions warfare where AI systems operate at speeds that transcend human reaction times, fundamentally altering the relationship between human commanders and military systems. According to Chinese researchers, future warfare will resemble "mythological fiction" with "Star Wars technology," suggesting their belief that AI will enable capabilities that seem almost magical by today's standards.

Military-Civil Fusion as Strategic Enabler

Perhaps no element of China's intelligentization strategy is more significant—or more concerning to competitors—than its comprehensive Military-Civil Fusion (MCF) approach. Under Xi Jinping's direct leadership, MCF has evolved from a bureaucratic coordination mechanism into what foreign analysts describe as a "whole-of-society" mobilization for technological supremacy.

The scale of this integration is unprecedented in modern history. According to the Foreign Policy Research Institute, China's MCF strategy "targets technologies such as quantum computing, semiconductors, 5G, nuclear technology, aerospace technology, gene editing and artificial intelligence to achieve military dominance." The 2017 establishment of the Central Commission for

Military-Civil Fusion Development, personally chaired by Xi Jinping, elevated MCF to the highest levels of state power, outranking traditional government ministries in China's hierarchical system.

This institutional elevation reflects the strategic logic underlying China's approach to military AI. Unlike the United States, where military and civilian AI development often proceed on parallel tracks with limited coordination, China's political system enables rapid and comprehensive integration of commercial innovations into military applications. As the National Bureau of Asian Research observes, MCF enables "two-for-one" benefits: "Beijing will not try to outspend the United States, but the objective remains the same: to build a system of policies and structures that will enable the PRC to rapidly modernize its military and more efficiently mobilize its economy in wartime."

The practical implementation of MCF creates what analysts describe as "innovation ecosystems" spanning military, academic, and commercial sectors. Major Chinese universities, including Beihang, Tsinghua, and the National University of Defense Technology, maintain extensive cooperation agreements with state-owned defense enterprises and private AI companies. According to recent reporting, these institutions receive "extensive funding from major defense state-owned enterprises, with whom they cooperate on UAV research, testing, and development."

The effectiveness of this approach is evident in China's rapid progress in military AI applications. The PLA's use of generative AI for military intelligence represents one concrete outcome of MCF integration. According to a January 2025 analysis by the Recorded Future's Insikt Group, "The PLA—which very likely rapidly adopted DeepSeek's generative AI models in early 2025—is likely using DeepSeek's LLMs for intelligence purposes, based on claims by a Chinese defense contractor that it has provided a DeepSeek-based OSINT model to the PLA."

Intelligentized Warfare Concepts: Redefining Combat

Information-Centric Approach to Dominance

China's conception of intelligentized warfare fundamentally reorients military strategy from kinetic destruction to information and cognitive dominance. This represents a sophisticated understanding of how AI can be weaponized not merely to improve existing military capabilities but to create entirely new forms of strategic power.

The foundational principle of China's approach is that "AI should be used kinetically and non-kinetically to dominate information systems to effectively paralyze an opponent's joint force." This concept extends far beyond traditional electronic warfare or cyber operations. According to Pentagon analysis, the PLA views warfare as "a confrontation between opposing operational systems, rather than annihilation of opposing mechanized military forces," leading to their emphasis on "systems destruction warfare as the next way of war."

This systems-oriented approach manifests in several key applications. The PLA is developing what it calls "network information systems" that use AI, cloud computing, and big-data techniques to fuse data from operational units and create "dynamic kill networks" across domains. As Defense

One reported in March 2025, these systems demonstrated their capabilities at the Zhuhai Air Show through products like Norinco's Intelligent Precision Strike System, which "used the UAVs' real-time data to model the battlefield, track targets, devise strike plans, distribute firing information, and execute follow-up strikes" with minimal human intervention.

The information-centric approach also emphasizes speed of decision-making as a critical advantage. Chinese researchers argue that AI enables "algorithmic warfare" where decision cycles are compressed from hours or days to seconds or minutes. This acceleration is not merely about faster targeting; it aims to achieve what Chinese strategists call "decision superiority"—the ability to observe, orient, decide, and act faster than opponents across all domains of conflict.

Cognitive Warfare Doctrine: Targeting the Human Mind

Perhaps the most distinctive aspect of China's intelligentized warfare doctrine is its explicit focus on cognitive warfare—the systematic use of AI, big data analytics, and psychological operations to manipulate enemy perceptions and decision-making. This represents an evolution of China's traditional "Three Warfares" doctrine (psychological, media, and legal warfare) enhanced by artificial intelligence automation.

The strategic logic of cognitive warfare rests on Sun Tzu's classical principle of subduing enemies without fighting. As Pentagon analysts note, "PLA researchers have stated that the victory of the cognitive narrative may yield greater strategic benefits than firepower destruction, force control, and siege, and that effects of CDO [Cognitive Domain Operations] can last long after the conflict has concluded."

Chinese military researchers have developed sophisticated frameworks for cognitive warfare that go far beyond traditional propaganda. According to the Hudson Institute's analysis, "Chinese strategists have stated that human cognition is the focus of intelligentized warfare, and that strategic objectives can be achieved through direct action on enemy cognition." This includes the deployment of "sophisticated machine learning models to create personalized propaganda aimed at Taiwan" and other targets.

The technological implementation of cognitive warfare leverages AI's capacity for personalization and scale. Rather than broadcasting generic propaganda, AI systems can analyze individual social media profiles, political preferences, and psychological vulnerabilities to craft targeted disinformation campaigns. The Washington Times reported in 2024 that Chinese military researchers are developing "sound weapons" and other neurostrike technologies designed to "induce confusion, emotional distress and a loss of consciousness" while directly impairing cognitive abilities.

The scope of cognitive warfare extends to what Chinese strategists call "mind dominance"—the use of propaganda as a weapon to influence public opinion and effect change in target nations' social systems. According to Pentagon assessments, the PLA views this as creating "an environment favorable to China and reduce civilian and military resistance to PLA actions." This represents warfare that begins long before any kinetic conflict, operating continuously in peacetime to shape the cognitive terrain.

Swarming and Autonomous Systems Integration

China's intelligentized warfare doctrine emphasizes the development of autonomous systems capable of operating in coordinated swarms that can overwhelm traditional defenses through sheer numbers and coordination. This approach reflects both China's manufacturing advantages and its strategic assessment that quantity, when properly coordinated by AI, can overcome qualitative disadvantages.

The "army of none" concept represents one manifestation of this strategy. According to Chinese military planning documents, the PLA is exploring the conversion of "thousands of retired J-7 and J-8 fighters into AI-enabled unmanned aircraft" to create massive swarms of autonomous platforms. This approach would leverage China's extensive inventory of older aircraft, transforming them from obsolete platforms into AI-enabled assets capable of saturating enemy defenses.

The technological foundation for swarm warfare rests on what Chinese researchers call "swarm learning," which enables "interconnected AIs to share learnings with each other to rapidly achieve higher capability." This distributed learning approach means that individual platforms within a swarm can benefit from the collective experience of all other platforms, creating adaptive networks that become more effective over time.

Recent Chinese military exercises have demonstrated increasingly sophisticated applications of autonomous swarms. According to Defense Connect's analysis of Chinese military developments, the PLA is developing "attritable, counter-stealth and persistent uncrewed aerial vehicles" that represent "a fresh approach to global combat" and "an open challenge to the traditional air warfare powers such as the United States of America and Russia."

The integration of swarm capabilities extends beyond individual platforms to encompass what Chinese strategists call "multi-domain precision warfare." This concept envisions coordinated operations across air, sea, land, space, cyber, and cognitive domains, with AI systems managing the complexity of simultaneous operations across multiple theaters. The ultimate vision is of warfare conducted by autonomous systems that can adapt, learn, and coordinate at scales and speeds impossible for human commanders to manage directly.

Practical Applications Analysis: From Theory to Reality

Advanced Systems Development and Deployment

China's transition from theoretical intelligentized warfare concepts to practical battlefield applications is evident in its rapid development and deployment of increasingly sophisticated AI-enabled military systems. The scope and pace of this development suggest that China is moving beyond experimental prototypes toward operational capabilities that could reshape regional military balances.

The People's Liberation Army Air Force has made particularly notable advances in unmanned systems integration. At the 2024 Airshow China, the PLA unveiled multiple new autonomous platforms, including a mysterious armed reconnaissance drone that Chinese military expert Song Zhongping described as "expected to be more powerful than the PLA Air Force's GJ-1 and GJ-2 armed reconnaissance drones." According to Global Times reporting, these new systems demonstrate capabilities for attacking "fixed, mobile and time-sensitive targets on the ground and on water surface" with enhanced autonomous decision-making.

Perhaps most significantly, China has operationalized several high-end unmanned systems that demonstrate the practical application of intelligentized warfare concepts. The WZ-8 high-speed, high-altitude reconnaissance drone, which reportedly operates at Mach 3 speed and 100,000 feet altitude, represents what Defense Connect describes as "the world's first hypersonic surveillance drone in operation." The WZ-7 Soaring Dragon and the redesigned GJ-11 stealth unmanned combat air vehicle have all achieved operational status, indicating that China has moved beyond developmental phases.

The Pentagon's 2024 assessment confirms this progression, noting that China's UAV modernization efforts are "rapidly matching U.S. standards" and that Beijing has achieved "several key milestones in the last three years" through airshow displays and operational appearances of new systems. This represents a significant acceleration from China's historical position as a follower in military aviation technology.

The integration of artificial intelligence into these platforms extends beyond simple automation. According to the Department of Defense analysis, "The PLA plans to use AI and machine learning to enhance missile sensors, which may make those missiles more accurate," while also pursuing "autonomous and precision-strike weapons" enabled by advanced AI capabilities.

Quantum Computing Integration: The Next Frontier

China's approach to intelligentized warfare increasingly incorporates quantum computing as a potential game-changing technology that could provide revolutionary advantages in processing speed and computational capability. This integration represents one of the most ambitious aspects of China's military modernization, with implications that extend far beyond conventional military applications.

Chinese investment in quantum technologies for military applications has reached unprecedented levels. According to MERICS analysis, "China sees quantum technology as pivotal in global science and technology competition and has stepped up government spending on scientific and industrial development to about USD 15 billion." The creation of the National Lab for Quantum Information Science in 2017, headed by China's leading quantum physicist Pan Jianwei, places quantum technology among the highest priority research areas.

The military applications of quantum computing in Chinese strategic thinking extend across multiple domains. The U.S. Defense Intelligence Agency's 2025 Worldwide Threat Assessment warns that "quantum technologies are nearing battlefield relevance, as rival nations invest in quantum sensors, communications and computing in ways that could erode American strategic advantages." For China, quantum technologies offer potential advantages in cryptography, communications security, and computational processing that could provide decisive military advantages.

In a groundbreaking development reported in April 2025, Chinese researchers successfully used the Origin Wukong quantum computer—featuring a domestically built 72-qubit chip—to fine-tune a billion-parameter artificial intelligence model. According to Global Times reporting, this represents "what they claim is a global first and a step toward combining quantum computing with advanced AI tasks." The experiment demonstrated that quantum hardware could improve AI model training performance even when traditional parameters were reduced, potentially offering solutions to the computational demands of military AI systems.

The strategic implications of quantum-AI integration are profound. Chinese researchers believe this convergence could enable "leapfrogging scientific breakthroughs" that provide asymmetric advantages over competitors relying on conventional computing architectures. As one Chinese quantum expert noted, the combination could enable "increases in calculating speed that are impossible to imagine," fundamentally altering the computational foundations of military decision-making.

Operational Testing and Real-World Validation

China's approach to validating its intelligentized warfare concepts increasingly relies on comprehensive testing that spans from computer simulations to real-world military exercises. This testing regime demonstrates both the sophistication of Chinese military AI development and the PLA's commitment to ensuring these systems perform effectively under operational conditions.

The PLA has established extensive simulation and training programs that incorporate AI-enabled scenarios. According to research from the Academy of Military Sciences, Chinese military researchers are using large language models and other AI technologies to "support intelligence analysis, strategic planning, simulation training, and command decision-making." These simulations enable the PLA to test intelligentized warfare concepts without the costs and risks associated with live-fire exercises.

Recent military exercises have demonstrated the operational integration of AI systems across multiple domains. The "Strait Thunder-2025A" exercise around Taiwan represented what The Diplomat analysis describes as "a qualitative shift—not merely larger in scale than previous 'Joint Sword' exercises, but operationally bolder in terms of deploying greater numbers of the PLA's major assets past the median line of the strait." These exercises incorporated autonomous systems and AI-enabled coordination at unprecedented scales.

The testing regime extends to emerging technologies that blur the lines between experimentation and operational deployment. Chinese researchers at the PLA National University of Defense Technology have published studies on applying generative AI to military operations, including the development of specialized military large language models trained on intelligence data from multiple sources including human intelligence, signals intelligence, and geospatial intelligence.

Perhaps most significantly, the PLA appears to be conducting continuous validation of AI systems through what amounts to persistent testing in contested environments. Japan's Ministry of Defense reported Chinese WZ-10 UAV flights over the East China Sea in May 2024, while Taiwan regularly reports Chinese drone patrols over claimed waters. These operations serve dual purposes: they advance Chinese territorial claims while providing real-world testing environments for autonomous systems and AI-enabled coordination.

Strategic Culture Impact: Authoritarian Advantages and Systemic Changes

Authoritarian Decision-Making as Technological Accelerator

China's authoritarian political system provides distinct advantages in the rapid development and deployment of military AI technologies, enabling decision-making speeds and resource mobilization that democratic systems often struggle to match. This structural advantage has become increasingly apparent as China accelerates its intelligentized warfare programs.

The concentration of decision-making authority in Xi Jinping's hands eliminates many of the bureaucratic obstacles that can slow technological adoption in democratic systems. As noted by analysts studying Chinese AI development, the "authoritarian decision-making structure" enables rapid integration of AI technologies across military and civilian sectors without the extended deliberation processes characteristic of democratic institutions. This centralized approach allows for swift pivots in strategic direction and massive resource allocation based on top-level decisions.

The effectiveness of this approach is evident in the pace of Chinese military AI advancement. According to the Center for Security and Emerging Technology analysis, while open-source information about specific PLA AI deployments remains limited, the breadth of research and development activities across seven major categories—including autonomous vehicles, surveillance and reconnaissance, and automated target recognition—demonstrates systematic and coordinated advancement across the entire military establishment.

China's willingness to accept technological risks that democratic societies might debate extensively provides another advantage. Chinese strategic culture has historically demonstrated acceptance of "risks, costs and long-term commitments for low-probability, high-return innovation." This risk tolerance enables rapid fielding of AI systems that might undergo extended testing and ethical review processes in democratic military establishments.

The authoritarian advantage extends to the integration of civilian and military AI development through the Military-Civil Fusion strategy. Unlike democratic systems where private companies maintain independence from military requirements, China's political system enables direct mobilization of civilian AI capabilities for military purposes. As noted by the Wire China analysis, this represents a fundamental difference from voluntary systems where "the private sector can opt out," versus China's "Leninist system where the party is above the law."

Long-term Strategic Thinking and Patient Capital

China's strategic culture emphasizes long-term planning horizons that provide significant advantages in developing transformative technologies like military AI. This approach contrasts sharply with the electoral cycles and budget approval processes that can constrain long-term technology development in democratic systems.

The temporal scope of Chinese strategic planning is evident in Xi Jinping's military modernization goals, which extend to 2049—the 100th anniversary of the People's Republic of China. This century-long planning horizon enables sustained investment in uncertain technologies that may not yield immediate returns but could provide revolutionary capabilities over decades. According to the Hudson Institute analysis, this long-term approach allows China to pursue "low-probability, high-return innovation" strategies that shorter-term focused systems might abandon.

China's approach to military AI development demonstrates this strategic patience through sustained investment across multiple technological pathways simultaneously. Rather than betting on single technological solutions, China pursues parallel development tracks in quantum computing, traditional semiconductors, alternative computing architectures, and advanced software systems. This portfolio approach increases the likelihood that at least some pathways will yield breakthrough capabilities.

The financial commitment underlying this long-term approach is substantial. While precise figures remain classified, estimates suggest that China's total national and local government spending on AI implementation reaches "the equivalent range of tens of billions of dollars," with at least two Chinese regional governments committing 100 billion yuan (~$14.7 billion) each to AI development. This represents sustained capital allocation at scales that few other nations can match.

Chinese strategic culture also demonstrates willingness to invest in fundamental research that may not yield immediate military applications. The establishment of the National Lab for Quantum Information Science and the integration of quantum research with AI development exemplify this approach. As noted by quantum technology analysts, China's quantum communication network—spanning 12,000 kilometers and including quantum satellites—demonstrates the practical results of sustained long-term investment.

Information Warfare and Narrative Control

China's strategic culture has evolved to prioritize information warfare and cognitive dominance as central elements of military strategy, viewing control over information flows and public perceptions as potentially more decisive than traditional kinetic capabilities. This cultural shift reflects both technological opportunities created by AI and fundamental beliefs about the nature of strategic competition.

The elevation of information warfare within Chinese strategic thinking is evident in the formal designation of "cognitive domain operations" as a core military capability. According to Pentagon analysis, the PLA defines cognitive domain operations as designed to achieve "mind dominance" through "the use of propaganda as a weapon to influence public opinion to effect change in a nation's social system." This represents warfare that operates continuously in peacetime, shaping the cognitive terrain long before any kinetic conflict.

China's approach to information warfare leverages AI's capacity for personalization and scale in ways that traditional propaganda methods cannot match. The deployment of "sophisticated machine learning models to create personalized propaganda aimed at Taiwan" demonstrates how AI enables targeting of specific individuals or groups with customized messaging designed to exploit particular psychological vulnerabilities or political preferences.

The strategic logic underlying this emphasis on cognitive warfare reflects Chinese beliefs about the changing character of conflict in the information age. As noted by War on the Rocks analysis, Chinese strategists believe that "intelligentized warfare" enables victory through direct manipulation of human cognition rather than traditional military defeat. This approach aims to achieve strategic objectives while avoiding the costs and risks associated with conventional military conflict.

The cultural acceptance of information warfare as a legitimate tool of statecraft provides China with operational advantages that democratic societies may struggle to match. While democratic militaries often face legal and ethical constraints on information operations against civilian populations, China's strategic culture faces fewer such limitations. This asymmetry enables more aggressive and comprehensive information warfare campaigns that can operate across multiple domains simultaneously.

The institutional infrastructure supporting information warfare has expanded significantly under Xi Jinping's leadership. The rebranding of the PLA Strategic Support Force as the Information Support Force in 2024 reflects the elevated priority placed on information operations. According to Asia Times analysis, this reorganization indicates "an evolution in Chinese strategic thought from 'winning informationized wars' to 'intelligentized warfare,' which in turn implies a broader mission set" encompassing psychological operations, cyber warfare, and cognitive manipulation.

Vulnerabilities and Constraints: Limitations of the Authoritarian Model

Technological Dependencies and Supply Chain Vulnerabilities

Despite China's rapid advancement in military AI, significant technological dependencies and supply chain vulnerabilities constrain the pace and scope of intelligentized warfare development. These limitations stem from China's continued reliance on foreign technologies in critical areas and the impact of Western export controls designed to limit Chinese access to advanced semiconductors and related technologies.

The semiconductor constraint represents the most significant limitation on China's military AI ambitions. According to CSIS analysis, the United States has implemented comprehensive export controls that exploit "chokepoint technologies in the global semiconductor technology supply chain" to limit Chinese capabilities. The October 2022 controls, tightened in October 2023 and December 2024, specifically target advanced semiconductors used for AI applications and the equipment necessary to manufacture them domestically.

The impact of these constraints is already visible in Chinese military AI development. While China has made progress in some areas, the Foreign Policy Research Institute notes that "advanced military power depends not on the superiority of the technology itself, but on the advanced operational theory of using the new technology." This suggests that technological limitations may force China to develop alternative operational concepts rather than directly matching Western capabilities.

China's response to semiconductor constraints reveals both adaptability and continued vulnerabilities. The evolution of Chinese semiconductor policy, as documented by the American Affairs Journal, shows Beijing implementing "China Semiconductor Industry Policy 3.0" with new top-down approaches and increased state control. However, as noted by industry analysts, "the future of China's semiconductor industry as of early 2024" remains uncertain, with "significant challenges" in moving beyond current technological nodes.

The smuggling and circumvention efforts documented by intelligence agencies indicate the severity of China's technological constraints. According to CSIS reporting, Chinese organizations circumvent restrictions by "using cloud platforms like Amazon Web Services and Oracle Cloud" to access cutting-edge AI chips without physically owning them. However, these workarounds provide limited access compared to unfettered ownership of advanced systems.

The constraint extends beyond semiconductors to other critical technologies. As noted by the U.S. Defense Intelligence Agency, "access to high-end microelectronics underpins a state's ability to develop computing-intensive advanced technologies, such as AI and quantum." China's difficulties in accessing advanced lithography equipment, electronic design automation software, and specialized materials create compounding limitations across the AI development pipeline.

Organizational and Cultural Limitations

China's authoritarian system, while providing certain advantages in rapid decision-making and resource mobilization, also creates organizational and cultural limitations that may impede the effective development and deployment of military AI systems.

The risk of "political distortions of military assessments" represents a fundamental challenge in authoritarian systems where reporting bad news can have career consequences. According to analysis of Chinese military culture, the pressure to provide optimistic assessments to senior leadership may lead to overconfidence in AI capabilities that have not been adequately tested under realistic conditions. This problem is particularly acute with AI systems, where performance can be highly dependent on training data and operational environments that may differ significantly from laboratory conditions.

The lack of real combat experience among PLA forces creates additional limitations in developing effective military AI systems. Unlike military forces that have tested technologies in actual combat, the PLA's AI development relies primarily on simulations and exercises. As noted by military analysts, this absence of combat validation means that Chinese AI systems may not perform as expected under the stress and unpredictability of actual warfare.

The brain drain affecting China's AI sector poses long-term constraints on military AI development. Despite substantial government investment, China continues to lose AI talent to international competitors offering better research environments and professional opportunities. According to industry reporting, sanctions and export controls have exacerbated this problem by limiting Chinese researchers' access to international collaboration and cutting-edge research tools.

The hierarchical nature of Chinese military culture may also limit the adaptability required for effective AI integration. Military AI systems often require rapid adaptation to changing battlefield conditions and the incorporation of lessons learned during operations. However, rigid hierarchies may slow the feedback loops necessary for continuous improvement of AI systems.

Quality control and validation represent additional organizational challenges. The rapid pace of Chinese AI development, driven by political pressure to demonstrate progress, may result in insufficiently tested systems being deployed prematurely. Unlike democratic systems where independent oversight and public scrutiny can identify problems, China's closed system may miss critical flaws until they manifest in operational settings.

International Isolation and Limited Testing Opportunities

China's increasingly isolated position in the international system creates constraints on its ability to validate and refine military AI technologies through international cooperation and real-world testing scenarios.

The limitation of alliance relationships restricts China's access to diverse testing environments and operational scenarios that could improve AI system robustness. Unlike Western military forces that regularly conduct joint exercises with multiple allies, providing varied testing conditions for AI systems, China's military AI development relies primarily on domestic testing and limited cooperation with a small number of partners.

The international response to China's military AI development has resulted in coordinated efforts to limit Chinese access to advanced technologies and research collaboration. According to allied assessments, these measures are "keeping quantum-related research on track" despite creating obstacles for Chinese development programs. The multilateral nature of these constraints means that China cannot easily circumvent limitations by working with alternative suppliers.

The lack of diverse operational testing environments limits Chinese ability to validate AI systems under varied conditions. While China conducts extensive exercises around Taiwan and in the South China Sea, these scenarios may not adequately test AI systems for operations in different geographic, climatic, or electromagnetic environments that could be encountered in actual conflicts.

The absence of meaningful military cooperation with advanced technological powers limits China's exposure to alternative approaches to military AI development. This isolation may result in the development of systems that are optimized for Chinese operational concepts but less effective against opponents using different strategies or technologies.

The international legal and ethical constraints on AI development create additional complications for Chinese military AI programs. While China may choose to ignore international norms regarding autonomous weapons systems, this approach could result in diplomatic costs and reduced effectiveness in scenarios requiring international cooperation or legitimacy.

Global Implications: Reshaping International Security

Challenge to Western Military-Technical Dominance

China's advancement in military AI technologies represents a fundamental challenge to the Western model of technological military superiority that has underpinned international security architecture since World War II. This challenge extends beyond simple capability competition to encompass different approaches to the development, deployment, and governance of military AI systems.

The scope of China's challenge is evident in its comprehensive approach to AI military integration. According to the Congressional Research Service, "China is widely viewed as the United States' closest competitor in the international AI market" with "recent Chinese achievements in the field demonstrating China's potential to realize its goals for AI development." This competition is not limited to specific platforms or systems but encompasses the entire technological ecosystem supporting military AI.

China's model of state-directed AI development presents an alternative to the market-driven approach characteristic of Western democracies. As noted by foreign policy analysts, China's ability to mobilize "every sector of society to contribute to the state's grand technological ambitions" through Military-Civil Fusion provides systemic advantages that purely market-based approaches may struggle to match. This model could prove attractive to other authoritarian states seeking to enhance their military capabilities.

The implications extend to alliance relationships and technology sharing arrangements. Western nations must now balance the benefits of AI cooperation with the risks that shared technologies could be acquired by China through various means. This creates tensions within alliance structures as different nations assess risks and opportunities differently.

The pace of Chinese AI military development is forcing Western militaries to accelerate their own programs, potentially with insufficient attention to safety and ethical considerations. The competitive pressure created by China's rapid advancement may lead to premature deployment of AI systems that have not been adequately tested or properly governed.

Influence on Other Authoritarian States

China's intelligentized warfare model is increasingly serving as a template for other authoritarian states seeking to enhance their military capabilities through AI integration. This diffusion of Chinese approaches to military AI development has significant implications for regional and global security.

The appeal of China's model lies partly in its promise of enabling smaller or technologically less advanced militaries to achieve asymmetric advantages over larger conventional forces. The emphasis on swarming tactics, cognitive warfare, and systems disruption offers pathways for states that cannot afford to match Western military capabilities platform-for-platform.

Russia's adoption of similar approaches demonstrates this pattern. According to analysis of Russian military AI development, Moscow is pursuing "asymmetric AI usage to overcome conventional disadvantages" using approaches that closely parallel Chinese concepts. The February 2024 agreement between China and Russia to "consult and coordinate on AI use in military technology" formalizes this cooperation and technology transfer.

Middle Eastern and other regional powers have shown increasing interest in Chinese military AI technologies and concepts. The export of Chinese military drones and AI-enabled systems provides practical experience with intelligentized warfare concepts while creating technological dependencies that may influence these nations' strategic orientations.

The normalization of AI-enabled autonomous weapons systems through Chinese development and deployment could accelerate their proliferation among other authoritarian states. Unlike democratic systems that may implement ethical constraints on autonomous weapons, authoritarian states may face fewer domestic obstacles to deploying increasingly capable AI military systems.

Impact on International Arms Control and Governance

China's approach to military AI development poses significant challenges to existing international arms control frameworks and efforts to develop governance mechanisms for emerging military technologies. The pace and scope of Chinese AI military development outstrip the ability of international institutions to develop appropriate regulatory responses.

The difficulty of verifying AI capabilities and development creates fundamental challenges for traditional arms control approaches. Unlike nuclear weapons or conventional military systems that can be observed and counted, AI military capabilities exist primarily as software and algorithms that are difficult to monitor or verify. This opacity makes it challenging to develop confidence-building measures or verification protocols.

China's rejection of certain international norms regarding autonomous weapons systems complicates efforts to develop global governance frameworks. While some nations advocate for preemptive restrictions on lethal autonomous weapons systems, China's continued development of increasingly capable autonomous systems suggests limited willingness to accept externally imposed constraints.

The dual-use nature of AI technologies makes traditional export control approaches less effective and creates challenges for alliance coordination. Technologies developed for civilian applications can often be rapidly adapted for military use, making it difficult to control the spread of militarily relevant capabilities while maintaining beneficial civilian cooperation.

The competitive dynamics created by Chinese AI military development may undermine international cooperation on AI safety and governance. As nations focus on maintaining military advantages, they may be less willing to share information or coordinate on safety research that could benefit potential competitors.

The success or failure of China's intelligentized warfare approach will likely influence global norms regarding the role of AI in military affairs. If Chinese AI military capabilities prove effective, other nations may feel compelled to adopt similar approaches regardless of their ethical or strategic preferences. Conversely, if Chinese systems prove unreliable or counterproductive, this could strengthen arguments for more restrictive approaches to military AI development.

The challenge extends to crisis management and escalation control in an era of AI-enabled military systems. Traditional concepts of deterrence and escalation management may prove inadequate when dealing with AI systems capable of operating at speeds that compress decision-making timelines and reduce opportunities for diplomatic intervention.

Conclusion: The Authoritarian AI Advantage

China's pursuit of intelligentized warfare represents more than military modernization; it embodies a comprehensive reimagining of how technology, strategy, and political organization can be integrated to achieve strategic advantage in the 21st century. The combination of authoritarian decision-making, long-term strategic planning, comprehensive Military-Civil Fusion, and willingness to accept technological risks has enabled China to achieve remarkable progress in military AI development despite significant constraints and limitations.

The global implications of this transformation extend far beyond the military realm. China's model demonstrates that authoritarian systems may possess certain structural advantages in developing and deploying transformative technologies, challenging assumptions about the relationship between political freedom and technological innovation. This has profound implications for how democratic societies approach their own AI development and the broader competition between different political and economic models.

However, the Chinese model also reveals significant vulnerabilities and constraints. Technological dependencies, organizational limitations, and international isolation create obstacles that may limit the ultimate effectiveness of intelligentized warfare concepts. The gap between theoretical capabilities and operational reality remains substantial, and the absence of combat validation means that many Chinese AI military systems remain untested under realistic conditions.

As China continues to advance its intelligentized warfare capabilities, the international community faces the challenge of developing appropriate responses that neither enable Chinese dominance nor sacrifice the values and principles that distinguish democratic societies. The choices made in the coming years regarding military AI development, international cooperation, and governance frameworks will likely determine whether artificial intelligence enhances or undermines international security and stability.

The Chinese approach to military AI represents both an opportunity and a warning. It demonstrates the transformative potential of artificial intelligence when systematically applied to military affairs, while also revealing the risks inherent in rapid technological development without adequate testing, validation, and international coordination. As the world's major powers continue to pursue their own versions of military AI, the lessons from China's experience—both positive and negative—will prove invaluable in shaping the future of warfare and international security.

Sources and References

1. **Center for a New American Security (CNAS)** - "Military Artificial Intelligence, the People's Liberation Army, and U.S.-China Strategic Competition" (February 1, 2024)
   - https://www.cnas.org/publications/congressional-testimony/military-artificial-intelligence-the-peoples-liberation-army-and-u-s-china-strategic-competition
2. **The International Affairs Review** - "China's Military Employment of Artificial Intelligence and Its Security Implications" (September 23, 2020)
   - https://www.iar-gwu.org/print-archive/blog-post-title-four-xgtap
3. **The Diplomat** - "Xi Jinping's Vision for Artificial Intelligence in the PLA" (November 16, 2022)
   - https://thediplomat.com/2022/11/xi-jinpings-vision-for-artificial-intelligence-in-the-pla/
4. **Hudson Institute** - "Xi Jinping's Vision for Artificial Intelligence in the PLA"
   - https://www.hudson.org/defense-strategy/xi-jinping-vision-artificial-intelligence-pla

5. **Recorded Future (Insikt Group)** - "China's PLA Leverages Generative AI for Military Intelligence" (1 month ago)
    - https://www.recordedfuture.com/research/artificial-eyes-generative-ai-chinas-military-intelligence
6. **Brookings Institution** - "The PLA's Strategic Support Force and AI Innovation" (January 3, 2023)
    - https://www.brookings.edu/articles/the-plas-strategic-support-force-and-ai-innovation-china-military-tech/
7. **Hudson Institute** - "Can China Build a World-Class Military Using Artificial Intelligence?"
    - https://www.hudson.org/defense-strategy/can-china-build-world-class-military-using-artificial-intelligence
8. **Center for Strategic and Budgetary Assessments (CSBA)** - "China's Ambitions for AI-Driven Future Warfare"
    - https://csbaonline.org/about/news/chinas-ambitions-for-ai-driven-future-warfare
9. **Center for Strategic and International Studies (CSIS)** - "China's Pursuit of Defense Technologies" (April 29, 2024)
    - https://www.csis.org/analysis/chinas-pursuit-defense-technologies-implications-us-and-multilateral-export-control-and
10. **The Wire China** - "China's Military-Civil Defusion" (September 30, 2024)
    - https://www.thewirechina.com/2024/09/22/chinas-military-civil-defusion/
11. **Foreign Policy Research Institute** - "China's Military-Civil Fusion Strategy: A Blueprint for Technological Superiority" (December 19, 2023)
    - https://www.fpri.org/article/2023/12/chinas-military-civil-fusion-strategy-a-blueprint-for-technological-superiority/
12. **National Bureau of Asian Research (NBR)** - "Commercialized Militarization: China's Military-Civil Fusion Strategy"
    - https://www.nbr.org/publication/commercialized-militarization-chinas-military-civil-fusion-strategy/
13. **MERICS** - "China's long view on quantum tech has the US and EU playing catch-up"
    - https://merics.org/en/report/chinas-long-view-quantum-tech-has-us-and-eu-playing-catch
14. **The Quantum Insider** - "Chinese Researchers Use Quantum Computer to Fine-Tune Billion-Parameter AI Model" (April 7, 2025)
    - https://thequantuminsider.com/2025/04/07/chinese-researchers-use-quantum-

computer-to-fine-tune-billion-parameter-ai-model/

15. **Asia Times** - "New PLA unit underscores intelligentized warfare shift" (April 26, 2024)
    - https://asiatimes.com/2024/04/new-pla-unit-underscores-intelligentized-warfare-shift/
16. **War on the Rocks** - "New Tech, New Concepts: China's Plans for AI and Cognitive Warfare" (April 15, 2022)
    - https://warontherocks.com/2022/04/new-tech-new-concepts-chinas-plans-for-ai-and-cognitive-warfare/
17. **Washington Times** - "Beijing designs military for war with U.S., including missiles, cyber-enabled psychological weapons" (January 10, 2025)
    - https://www.washingtontimes.com/news/2024/dec/18/chinese-military-adds-hundreds-missiles-arsenal-pe/
18. **The Diplomat** - "China's Chilling Cognitive Warfare Plans" (May 7, 2024)
    - https://thediplomat.com/2024/05/chinas-chilling-cognitive-warfare-plans/
19. **DefenseScoop** - "Pentagon report warns that China may wield AI to enhance 'cognitive domain operations'" (November 30, 2022)
    - https://defensescoop.com/2022/11/30/pentagon-report-warns-that-china-may-wield-ai-to-enhance-cognitive-domain-operations/
20. **Defense One** - "New products show China's quest to automate battle" (March 2, 2025)
    - https://www.defenseone.com/threats/2025/03/new-products-show-chinas-quest-automate-battle/403387/
21. **DefenseScoop** - "China's drone modernization efforts close to 'matching US standards,' Pentagon report says" (December 18, 2024)
    - https://defensescoop.com/2024/12/18/chinas-drone-modernization-efforts-close-to-matching-us-standards-pentagon-report-says/
22. **DRONELIFE** - "China's Shift to Small Drones Signals Reshaping of Modern Warfare" (April 30, 2025)
    - https://dronelife.com/2025/04/30/chinas-shift-to-small-drones-reshapes-modern-warfare/
23. **Defense One** - "China's burgeoning drone arsenal shows power of civil-military fusion" (1 month ago)
    - https://www.defenseone.com/ideas/2025/06/chinas-drone-arsenal-shows-power-civil-military-fusion/406118/
24. **The Diplomat** - "The Drone Swarm Paradox in the Taiwan Strait" (3 weeks ago)
    - https://thediplomat.com/2025/06/the-drone-swarm-paradox-in-the-taiwan-strait/

25. **CSIS** - "Choking off China's Access to the Future of AI" (October 15, 2024)
    - https://www.csis.org/analysis/choking-chinas-access-future-ai
26. **CSIS** - "The Limits of Chip Export Controls in Meeting the China Challenge" (May 7, 2025)
    - https://www.csis.org/analysis/limits-chip-export-controls-meeting-china-challenge
27. **American Affairs Journal** - "A New Era for the Chinese Semiconductor Industry: Beijing Responds to Export Controls" (February 23, 2024)
    - https://americanaffairsjournal.org/2024/02/a-new-era-for-the-chinese-semiconductor-industry-beijing-responds-to-export-controls/

Chapter 6: Russia's Digital Asymmetry: AI as the Great Equalizer

Surviving Through Innovation

Opening: The Shahed Evolution

In December 2024, Ukrainian military sources confirmed a dramatic shift in Russia's approach to autonomous warfare. Russian forces had begun equipping their Shahed-136 strike drones with artificial intelligence systems, making these weapons "smarter" and better able to evade Ukrainian air defenses while identifying high-value targets such as energy infrastructure. According to Ukrainian officials, these upgraded drones no longer relied solely on GPS navigation but incorporated machine vision algorithms that compared onboard camera images to pre-loaded terrain maps in real-time.

What made this development particularly significant was not just the technical upgrade, but what it represented: Russia's systematic transformation from conventional military superiority to AI-enabled asymmetric warfare. The evolution was documented by Ukrainian officials who observed that newer Shahed variants, labeled "MS 236," now featured electro-optical guidance heads and thermal imaging cameras. A January 2024 leak of classified documents revealed these upgrades, which were publicly displayed during Vladimir Putin's visit to the Special Technology Center in St. Petersburg in March 2024.

Even more advanced models were reportedly under development. According to Ukrainian Major General Vladyslav Klochkov, Russian forces were field-testing the Shahed MS001 powered by Nvidia Jetson Orin processors, which he described as "an autonomous combat platform that sees, analyzes, decides, and strikes without external commands."

This technological leap was no accident. In late 2024, Russian President Vladimir Putin had announced plans for "autonomous robots with AI to be used massively from 2025-2026," signaling a fundamental shift in Russian military strategy. For a nation that had suffered significant conventional military setbacks in Ukraine, artificial intelligence represented not just a technological upgrade but an existential necessity—a way to offset asymmetries in conventional power through innovation.

Strategic Imperative Analysis

Asymmetric Response to Conventional Weakness

Russia's turn to AI-enabled military systems stems from a stark strategic reality: the need to compensate for massive conventional military losses in Ukraine while facing a widening capability gap with Western forces. According to the Royal United Services Institute, Russia is using AI to "offset asymmetry in conventional power due to extensive personnel and material losses in Ukraine."

This strategic imperative is reflected in Russia's dramatic military budget increases. According to the Stockholm International Peace Research Institute (SIPRI), Russia's military expenditure reached an estimated $149 billion in 2024, representing a 38% increase from 2023 and equivalent to 7.1% of Russia's GDP. For 2025, Russia's total planned military expenditure can be estimated at 15.5 trillion rubles, a real-terms increase of 3.4% over 2024 and equivalent to 7.2% of gross domestic product.

When adjusted for purchasing power parity (PPP), Russia's defense spending tells an even more striking story. According to the International Institute for Strategic Studies (IISS), Russia spent nearly $462 billion in 2024 when adjusted for PPP, effectively surpassing Europe's collective defense spending. This level of military investment represents the highest as a percentage of GDP since the Soviet era.

Putin has consistently framed this technological development in existential terms. Russian authorities believe that "trends such as the proliferation of autonomous and AI-enabled weapons systems and the convergence between human-machine learning, cyber and AI, coupled with novel operational concepts and force structures, will change the trajectory and character of future warfare." As Putin has stated, technological development represents a matter of survival: "one either succeeds or they will be annihilated."

Strategic Logic of Technological Leapfrogging

Russia's AI strategy represents a calculated bet on technological leapfrogging—using advanced technologies to skip generations of conventional military development. The logic follows that deployed by economically weaker powers throughout history: rather than matching adversaries' strengths directly, leverage emerging technologies to create new forms of asymmetric advantage.

Putin emphasized this approach in his 2024 remarks at the Valdai Discussion Club, where he coined the term "sovereign artificial intelligence." He declared that "Russia must develop its own sovereign artificial intelligence based on national algorithms and laws" and positioned this as essential for competing with global AI leaders. The Russian leadership views AI not just as a military tool but as a pathway to maintaining great power status despite economic constraints.

This technological sovereignty strategy is driven by both opportunity and necessity. Western sanctions have limited Russia's access to advanced semiconductors and AI components, forcing Moscow to develop domestic alternatives while seeking partnerships with countries like China. As Putin noted in November 2024, "Competition is intensifying. I'm not speaking of armed confrontation, but economic rivalry makes AI development inevitable."

The pressure for battlefield results has accelerated development timelines. Putin has referred to experiences from the battlefield in Ukraine, noting that "the most effective weapons systems are those that operate quickly and 'almost in an automatic mode.'" This battlefield feedback loop has created unprecedented urgency in Russian AI development programs.

"Sovereign Artificial Intelligence" Strategy

Conceptual Framework

Russia's approach to AI development centers on the concept of "sovereign artificial intelligence"—a term coined by Putin that gained prominence after his remarks at the Valdai Club meeting in November 2024. This concept encompasses both technological independence and ideological control over AI systems.

According to T-invariant analysis, Putin's vision of sovereign AI emerged from his belief that "these platforms are mostly developed abroad, and they shape worldviews." The response, as Putin articulated, is clear: "Here, we must recognize this and develop our own sovereign artificial intelligence. Of course, we should use everything available, but we must also advance our own initiatives in this field."

The concept reflects Russia's broader strategic culture of technological sovereignty, similar to previous initiatives in nuclear technology and space capabilities. However, the AI implementation has unique characteristics driven by wartime pressures and immediate battlefield requirements.

Institutional Development

Russia has established a comprehensive institutional framework for military AI development. The Ministry of Defense created a special AI department in 2021, headed by Vasilii Yelistratov, which coordinates AI development across the Russian military. This department works alongside the Advanced Research Foundation (FPI), established in 2012 as Russia's equivalent to DARPA.

The ERA Technopolis, established in 2018 in Anapa, serves as Russia's premier military AI development center. The facility brings together over 100 entities including major arms manufacturers like Kalashnikov and Sukhoi, civilian universities, and research institutions. Eight specialized "scientific companies" (nauchnye roty) operate within ERA Technopolis, working on specific AI applications for different military branches.

A crucial development has been the integration of academic and military research. Moscow State University, led by Putin's daughter Katerina Tikhonova at its AI Institute, exemplifies this approach. As Tikhonova acknowledged, Russia is currently "following trends rather than setting them" in AI development, but argues that the country has a "more pragmatic approach to AI development" shaped by specific national challenges, particularly "applications in UAVs (unmanned aerial vehicles) and the oil and gas sector."

Ukraine War Applications

EW-Proof Autonomous Systems

The Ukraine conflict has served as both testing ground and accelerator for Russian AI military applications. One of the most significant developments has been the creation of systems designed to overcome Ukrainian electronic warfare capabilities.

Russian forces have developed autonomous systems that are "EW-proof due to AI's capability to process electronic threats and operate independently." These systems represent a direct adaptation to Ukrainian electronic warfare tactics that successfully disrupted earlier Russian drone operations through jamming and GPS spoofing.

The technical approach involves AI models that can:

- Process and react to electronic threats autonomously
- Navigate without GPS using computer vision and terrain mapping
- Adapt dynamically to changing electronic warfare environments
- Continue missions despite communication link disruption

According to Ukrainian sources, Russia has deployed the KUB-LA kamikaze drone, produced by ZALA Aero Group (a subsidiary of Kalashnikov), which claims to be able to select and engage targets using AI technology. Another example is the Lancet-3 loitering munition, which according to its producer is highly autonomous due to sensors that enable it to locate and destroy targets without human guidance.

Organizational Transformation

The war has prompted significant organizational changes in Russian military AI development. Putin announced the creation of a new military branch dedicated to unmanned systems, marking a transition from what experts call the "guerrilla era of ad hoc drone warfare" to systematic institutional development.

This transformation includes:

Dedicated Military Structures: The establishment of specialized units for unmanned systems development and deployment, moving beyond improvised volunteer efforts to professional military capabilities.

Accelerated Development Cycles: The pressure for "immediate battlefield deployment" has compressed traditional military development timelines, forcing rapid iteration and testing of AI systems in combat conditions.

Lesson Integration: The Ministry of Defense launched the Rubicon Center in August 2024 specifically to "systematize lessons from Ukraine, including the development and application of AI."

Technology Under Fire

Combat requirements have accelerated Russian AI development in ways peacetime programs could not achieve. The war creates what defense analysts describe as "pressure for immediate battlefield deployment" that forces rapid advancement of AI technologies.

Key developments include:

Real-Time Adaptation: Russian forces now adapt to Ukrainian innovations in "two or three days" versus previous month-long cycles, demonstrating the accelerated pace of AI-enabled military adaptation.

Combat Testing: Russian AI systems are being tested and refined under actual battlefield conditions, providing real-world data that laboratory environments cannot replicate.

Production Scaling: According to Putin's April 2025 statements, frontline units are receiving around 4,000 First-Person View (FPV) drones daily, indicating massive scaling of production capabilities.

Hybrid Warfare Integration

Multi-Domain AI Applications

Russia is pursuing AI integration across multiple domains simultaneously, reflecting its doctrine of hybrid warfare that blurs lines between conventional and unconventional conflict. According to the Center for Naval Analyses, Russia is applying AI "in information and cyber operations" alongside kinetic capabilities, and is "likely to apply AI in nuclear forces command, control, management, and use."

Electronic Warfare Evolution: Russian forces are developing AI-enabled systems designed to be immune to electronic warfare through autonomous threat processing. This represents a direct response to Ukrainian electronic warfare successes that have disrupted traditional drone operations.

Information Operations Enhancement: Russia has deployed AI to scale its information warfare capabilities. According to joint statements by the FBI and allied agencies, "RT affiliates used Meliorator – a hidden software package enhanced by artificial intelligence (AI) – to create fictitious online users" to spread disinformation across social networks.

Cyber-AI Convergence: Russian military thinking emphasizes the "convergence between human-machine learning, cyber and AI" as fundamentally changing warfare character. This includes using AI to automate cyberattacks and enhance the effectiveness of traditional "active measures" information operations.

Dual-Use Integration Philosophy

Russia has developed a strategic culture that seamlessly integrates military and civilian AI development, viewing technological advancement as serving both economic and security objectives simultaneously.

According to German Gref, CEO of Sberbank and a key figure in Russian AI development, the introduction of AI could increase Russia's GDP by 1% by 2025. Putin and other central players anticipate that defense innovation—and AI in particular—will produce dual-use technology that can drive nationwide economic growth, which is needed now more than ever given the weight of Western sanctions.

This dual-use approach extends to international cooperation. On December 30, 2024, Putin ordered the Russian government and Sberbank to collaborate with China on artificial intelligence research, with Russian Prime Minister Mikhail Mishustin and Sberbank CEO Herman Gref required to submit a roadmap for collaboration by April 30, 2025.

Strategic Culture Evolution

From Brute Force to Asymmetric Intelligence

Russia's traditional strategic culture emphasized overwhelming force and material superiority—a legacy of Soviet military doctrine that relied on mass mobilization and numerical advantage. The Ukraine war has forced a fundamental reconsideration of this approach.

The shift toward AI represents a cultural adaptation from quantitative to qualitative advantages. As Russian Defense Minister Andrei Belousov stated in October 2024, "AI-powered drones are playing a pivotal role on the battlefield," reflecting the military leadership's recognition that technological sophistication can substitute for numerical superiority.

This cultural shift manifests in several ways:

Innovation Over Production: Rather than simply manufacturing more conventional weapons, Russian strategy now emphasizes developing smarter systems that can achieve greater effects with fewer resources.

Rapid Experimentation: The traditional Soviet preference for extensive testing and gradual deployment has given way to rapid fielding of experimental systems under combat conditions.

Technological Optimism: Russian military culture has embraced the potential for AI to provide decisive advantages, despite the technology's current limitations and uncertainties.

Rapid Adaptation Culture

The Ukraine war has accelerated the development of a strategic culture that prizes rapid adaptation and continuous learning. Russian forces now demonstrate adaptation cycles measured in days rather than months, a dramatic departure from traditional military culture.

This adaptation culture is exemplified by Putin's April 2025 statements emphasizing that "the experience gained during the ongoing military operations is driving innovation in Russian military technology." He noted "real breakthroughs in military equipment" with some developments being "unique" and "first of their kind globally."

The adaptation culture includes:

Compressed Decision Cycles: Military leaders are making technology deployment decisions in weeks rather than years, accepting higher risks in exchange for faster capability development.

Learning Organization: The establishment of institutions like the Rubicon Center demonstrates systematic efforts to capture and apply combat lessons to technology development.

Experimental Mindset: Russian forces have shown willingness to deploy untested AI systems in combat, using battlefield results to iterate and improve capabilities.

Constraints and Challenges

Sanctions and Technology Access

Western sanctions have created significant obstacles for Russian AI military development, particularly in accessing advanced semiconductors and specialized components.

Hardware Limitations: Sberbank CEO German Gref acknowledged in 2023 that replacing graphics processing units (GPUs)—critical for AI development—has proven "particularly challenging" for Russia. Sanctions have forced Russian developers to rely on lower-performance alternatives or complex workaround solutions.

Component Availability: As noted by T-invariant analysis, "establishing stable manufacturing becomes impossible when components are available today, gone tomorrow, and then reappear the day after as technically identical yet different batches." While Ukraine can acquire Nvidia Jetson processors for autonomous robotics as "merely a financial challenge," Russia faces "far greater constraints."

Innovation Ecosystem Disruption: The departure of thousands of IT specialists from Russia has damaged the AI sector's human capital base, while the rupture of international cooperation agreements has limited access to global research networks.

Structural Economic Problems

Russia's AI development operates within broader economic constraints that limit long-term sustainability and scalability.

Resource Competition: The massive military spending increases—with defense consuming over 40% of the total budget—come at the expense of civilian investment, education, and infrastructure development that support technological innovation.

Brain Drain: The emigration of technical talent has reduced the domestic workforce available for AI development, while international isolation limits Russia's ability to attract foreign expertise.

Industrial Base Limitations: Despite import substitution efforts, Russia's domestic technology manufacturing capabilities remain limited compared to global leaders, constraining the scale and sophistication of AI system production.

Verification and Reliability Challenges

The pressure for rapid deployment under combat conditions creates risks of deploying insufficiently tested AI systems with potential strategic consequences.

Limited Testing: Unlike peacetime development programs that allow extensive testing and validation, wartime pressures encourage deployment of systems before their limitations are fully understood.

Human-AI Integration: Russian military culture's traditional emphasis on centralized control conflicts with the distributed decision-making that AI systems enable, creating organizational adaptation challenges.

Strategic Miscalculation: Over-reliance on untested AI capabilities could lead to strategic misjudgments, particularly given Russia's limited recent experience with large-scale conventional warfare prior to Ukraine.

International Cooperation and Partnerships

China-Russia AI Collaboration

Russia's most significant international AI partnership is with China. The collaboration gained new momentum with Putin's December 2024 directive ordering enhanced cooperation between Russia and China on AI research and development.

This partnership addresses several Russian constraints:

Technology Access: China can provide access to advanced semiconductors and AI components that Western sanctions have denied Russia.

Research Collaboration: Joint development programs leverage Chinese AI expertise while providing China with real-world military testing data from Ukraine.

Standards Development: The partnership aims to create alternative AI standards and frameworks to compete with Western-dominated systems.

BRICS and Broader Partnerships

Russia has also announced plans for an AI alliance within BRICS (Brazil, Russia, India, China, South Africa), extending AI cooperation beyond bilateral arrangements with China. Additionally, Russia and Iran signed a memorandum of understanding on AI development in March 2024 to enhance "technological and cultural dialogue" and conduct research on AI applications.

These partnerships serve multiple purposes:

Sanctions Evasion: Diversifying technology sources and supply chains to reduce vulnerability to Western restrictions.

Market Development: Creating alternative markets for Russian AI technologies and applications.

Norm Setting: Establishing alternative frameworks for AI governance and military applications that compete with Western standards.

Future Implications

Model for Resource-Constrained Powers

Russia's approach to AI military development may provide a template for other nations facing similar resource constraints and technology access limitations.

The Russian model demonstrates several potentially replicable strategies:

Asymmetric Focus: Concentrating limited resources on specific AI applications that provide maximum strategic advantage rather than attempting to match leaders across all domains.

Rapid Fielding: Accepting higher risks to deploy AI systems quickly, using combat experience to drive iterative improvement.

State Coordination: Employing centralized direction to maximize limited resources and ensure civilian and military AI development serve complementary objectives.

Partnership Strategy: Leveraging relationships with technology partners to access capabilities and expertise not available domestically.

Regional Influence and Technology Transfer

Russia's AI military developments are likely to influence other nations in its sphere of influence and among its strategic partners.

Technology Sharing: Russia's agreements with countries like China and Iran to share AI military technology could accelerate autonomous weapons proliferation beyond current major powers.

Doctrine Export: Russian concepts like "sovereign AI" and hybrid warfare integration may influence other nations' approaches to military AI development.

Regional Competition: Russia's AI capabilities may prompt regional rivals to accelerate their own AI military programs, contributing to arms race dynamics.

Long-term Strategic Competition

Russia's AI strategy reflects a long-term vision of strategic competition where technological sovereignty becomes essential for maintaining great power status.

Persistent Challenge: Despite current limitations, Russia's systematic approach to AI development suggests it will remain a significant competitor in military AI, particularly in specific niche areas.

Innovation Under Pressure: The combination of resource constraints and strategic pressure may drive Russian innovation in ways that peaceful competitors cannot replicate, potentially yielding unexpected breakthroughs.

Alliance Implications: Russia's growing technological cooperation with China and other partners creates new forms of strategic competition that may challenge Western assumptions about technological leadership and control.

Conclusion: Digital Resilience Through Innovation

Russia's approach to military AI represents a case study in strategic adaptation under pressure. Faced with conventional military setbacks, economic sanctions, and technological isolation, Russia has pursued an AI strategy designed to leverage asymmetric advantages while building technological sovereignty.

The Russian model demonstrates both the potential and limitations of state-directed AI development under resource constraints. While sanctions and economic pressures limit Russia's ability to match Western and Chinese AI capabilities across all domains, the focused approach to military applications and the pressure of combat conditions have accelerated development in specific areas.

Perhaps most significantly, Russia's experience illustrates how AI can serve as an equalizer for powers that lack conventional military superiority. By developing AI systems that can operate in contested environments, overcome electronic warfare, and coordinate autonomous operations, Russia is creating capabilities that could challenge larger, better-equipped forces.

The long-term implications extend beyond Russia itself. The Russian model of AI development—emphasizing rapid deployment, state coordination, asymmetric applications, and international partnerships—may become attractive to other nations facing similar strategic challenges. As AI technologies continue to proliferate and costs decrease, Russia's approach to digital asymmetry may provide a template for how medium and smaller powers can leverage artificial intelligence to maintain strategic relevance in an era of great power competition.

The question facing Western strategists is not whether Russia will achieve AI military parity with leading powers, but whether Russian innovations in AI-enabled asymmetric warfare will create new forms of strategic challenge that require fundamental adaptations in Western military thinking and capability development. Russia's digital asymmetry strategy may represent the first systematic attempt by a major power to use AI as a great equalizer—and its success or failure will likely influence how other nations approach military AI in an increasingly competitive international environment.

Sources and References

1. Breaking Defense. "Inside Russia's plan to build autonomous drone swarms." January 8, 2025. https://breakingdefense.com/2025/01/inside-russias-plan-to-build-autonomous-drone-swarms/
2. Tom's Hardware. "Russia allegedly field-testing deadly next-gen AI drone powered by Nvidia Jetson Orin." January 2025. https://www.tomshardware.com/tech-industry/artificial-intelligence/russia-allegedly-field-testing-deadly-next-gen-ai-drone-powered-by-nvidia-jetson-orin-ukrainian-military-official-says-shahed-ms001-is-a-digital-predator-that-identifies-targets-on-its-own
3. Royal United Services Institute. "Struggling, Not Crumbling: Russian Defence AI in a Time of War." 2024. https://www.rusi.org/explore-our-research/publications/commentary/struggling-not-crumbling-russian-defence-ai-time-war
4. Stockholm International Peace Research Institute. "Unprecedented rise in global military expenditure as European and Middle East spending surges." April 28, 2025. https://www.sipri.org/media/press-release/2025/unprecedented-rise-global-military-expenditure-european-and-middle-east-spending-surges
5. Stockholm International Peace Research Institute. "Preparing for a Fourth Year of War: Military Spending in Russia's Budget for 2025." 2025. https://www.sipri.org/publications/2025/sipri-insights-peace-and-security/preparing-fourth-year-war-military-spending-russias-budget-2025
6. Breaking Defense. "Russia overtakes all of Europe on defense spending in key metric: IISS military balance." February 12, 2025. https://breakingdefense.com/2025/02/russia-overtakes-all-of-europe-on-defense-spending-in-key-metric-iiss-military-balance/

7. Center for a New American Security. "The Role of AI in Russia's Confrontation with the West." September 9, 2024. https://www.cnas.org/publications/reports/the-role-of-ai-in-russias-confrontation-with-the-west
8. T-invariant. "Sovereign Means Military: How Russia Militarized AI, Drone, and Cryptography Industries." March 2025. https://t-invariant.org/2025/03/sovereign-means-military-how-russia-militarized-ai-drone-and-cryptography-industries/
9. Chatham House. "Advanced military technology in Russia: Military applications of artificial intelligence." September 2021. https://www.chathamhouse.org/2021/09/advanced-military-technology-russia/06-military-applications-artificial-intelligence
10. Center for Naval Analyses. "AI and Autonomy in Russia." https://www.cna.org/our-media/newsletters/ai-and-autonomy-in-russia
11. Defence Blog. "Putin eyes AI and lasers for future warfare." April 25, 2025. https://defence-blog.com/putin-eyes-ai-and-lasers-for-future-warfare/
12. The Washington Quarterly. "Russia's Drive for AI: Do Deeds Match the Words?" Vol. 47, No. 4, 2024. https://www.tandfonline.com/doi/full/10.1080/0163660X.2024.2435162
13. Jamestown Foundation. "Russia Releases Proposed Military Budget for 2025." October 9, 2024. https://jamestown.org/program/russia-releases-proposed-military-budget-for-2025/
14. OSW Centre for Eastern Studies. "Russia's budget for 2025: war above all." December 19, 2024. https://www.osw.waw.pl/en/publikacje/osw-commentary/2024-11-22/russias-budget-2025-war-above-all
15. Wikipedia. "Military budget of Russia." Updated weekly. https://en.wikipedia.org/wiki/Military_budget_of_Russia
16. Wikipedia. "Artificial intelligence arms race." Updated weekly. https://en.wikipedia.org/wiki/Artificial_intelligence_arms_race
17. RealClearDefense. "Russia Capitalizes on Development of AI in its Military Strategy." March 4, 2025. https://www.realcleardefense.com/2025/03/04/russia_capitalizes_on_development_of_ai_in_its_military_strategy_1095190.html
18. MacroTrends. "Russia Military Spending/Defense Budget | Historical Chart & Data." https://www.macrotrends.net/global-metrics/countries/rus/russia/military-spending-defense-budget

Chapter 7: Middle Powers' AI Strategies: Innovation Amidst Competition

Finding Niche Advantages in the AI Ecosystem

Opening Scene: The Alliance Dilemma

The conference room at NATO headquarters in Brussels buzzed with tension as defense ministers from across the Alliance grappled with a fundamental challenge of the AI age. On the screens before them, charts displayed the sobering reality: artificial intelligence systems developed by different nations were increasingly sophisticated, yet potentially incompatible with one another. As one senior NATO official would later observe, there was a genuine risk that "AI systems developed by different nations may contradict each other and slow down the targeting cycle rather than speeding it up."

The July 2024 NATO Summit had just endorsed the Alliance's revised artificial intelligence strategy, recognizing that "future high-speed conflicts require sharing large volumes of AI-parsed data to connect sensors to shooters." Yet the fundamental tension remained: how could middle powers maintain their national advantages in AI while ensuring their systems could operate seamlessly with allies? This dilemma would define the strategic choices facing nations across Europe, the Asia-Pacific, and beyond as they navigated between competition and cooperation in the AI age.

Europe's Collective AI Approach

NATO's Updated Strategy: Building Alliance-Wide Capabilities

The transformation of NATO's approach to artificial intelligence represents one of the most significant adaptations in the Alliance's 75-year history. Building on the foundation established in 2021, NATO released its revised AI strategy in July 2024, explicitly addressing the challenges posed by "generative AI and AI-enabled information tools" that had emerged with unprecedented speed and capability.

The 2024 strategy fundamentally reimagined how the Alliance would approach military AI, establishing six core Principles of Responsible Use (PRUs): Lawfulness, Responsibility and Accountability, Explainability and Traceability, Reliability, Governability, and Bias Mitigation. These principles represent more than aspirational goals—they form the foundation for NATO's approach to ensuring that AI systems "throughout the Alliance" can achieve genuine interoperability.

Central to this effort is NATO's Defence Innovation Accelerator for the North Atlantic (DIANA), which has rapidly expanded its footprint across the Alliance. As of 2024, DIANA operates through a network of more than 200 accelerator sites and test centers across 28 Allied countries, representing a doubling of capacity since its initial establishment. This network enables

NATO to support innovators "from across the Alliance as they develop their technologies," creating pathways for rapid technology adoption while maintaining Alliance-wide standards.

The €1 billion NATO Innovation Fund, established as "the world's first multi-sovereign venture capital fund," represents an unprecedented experiment in collective technology investment. With 24 NATO Allies participating as Limited Partners, the fund explicitly prioritizes "investments in companies accelerated through DIANA" while ensuring that technological advances benefit the entire Alliance rather than individual nations alone.

Perhaps most significantly, NATO's 2024 strategy identifies emerging threats that had not been fully appreciated in earlier iterations. For the first time, the strategy explicitly addresses "AI-enabled disinformation, information operations and gender-based violence as issues of concern for the Alliance." This expansion reflects growing recognition that military AI encompasses far more than traditional weapons systems, extending into information warfare and societal resilience.

The strategy's emphasis on developing "Alliance-wide AI Testing, Evaluation, Verification & Validation (TEV&V) landscape" through DIANA's network represents a practical response to the interoperability challenge. As NATO officials recognize, autonomous systems must be "programmed to operate together without relying on common sense or relationships between human operators." This requires unprecedented technical coordination and shared standards development across sovereign nations with distinct technological approaches.

EU's Regulatory Leadership: Setting Global Standards

The European Union has chosen a fundamentally different path in AI governance, one that prioritizes regulatory leadership over competitive advantage. The EU AI Act, which entered into force in 2024, represents the world's first comprehensive regulation of artificial intelligence, establishing a "human-centric, risk-based model" that contrasts sharply with the "US flexible, innovation-friendly standards."

This regulatory approach reflects a broader EU strategy of achieving "digital sovereignty" through "jurisdictional independence" rather than direct technological competition with the United States and China. By establishing clear rules for AI development and deployment, the EU seeks to shape global norms while ensuring that European values of privacy, human rights, and democratic governance are embedded in AI systems.

The AI Act's exclusion of military applications creates both opportunities and challenges for European defense establishments. While Article 2 explicitly states that the regulation "does not apply to AI systems used for military, defense, or national security purposes," the dual-use nature of most AI technologies means that civilian regulations will inevitably influence military applications.

European defense innovation has proceeded through multiple channels despite this regulatory gap. The European Defence Fund has allocated "almost EUR 8 billion for 2021-2027" to advanced technologies including AI, while "multiple European Defence Fund and PESCO projects" focus specifically on "integrating AI into future military capabilities." This represents a significant commitment to ensuring that Europe develops indigenous AI capabilities for defense applications.

The EU's approach reflects a broader strategic logic: rather than competing directly with superpowers in raw AI capability, Europe seeks to establish the rules and standards by which AI systems operate globally. As analysts note, this represents an attempt to ensure that "whoever sets the AI rules exports their worldview," positioning Europe as the regulatory authority in an increasingly contested technological domain.

Coordination Challenges: The Price of Collective Action

The practical challenges of coordinating AI systems across multiple sovereign nations have proven more complex than initially anticipated. NATO's recognition that "computer vision systems from different nations" might "contradict each other and slow down targeting cycle rather than speeding it up" reflects the fundamental tension between national technological sovereignty and collective military effectiveness.

Technical interoperability represents only the first layer of these challenges. Each NATO member nation has developed distinct approaches to AI integration based on their unique strategic cultures, regulatory environments, and technological capabilities. Ensuring that these diverse systems can operate together requires not just technical standards but shared operational concepts and decision-making frameworks.

The classification challenges identified by NATO officials add another layer of complexity. As one senior defense official noted, "no one's come up with great answers for appropriate classification of AI products," particularly when dealing with "classification by compilation" where AI systems can "piece together classified information from unclassified sources." This challenge becomes exponentially more complex when multiple nations must share AI-processed intelligence while maintaining operational security.

NATO's response has been to establish formal frameworks for coordination. The Alliance's Data and Artificial Intelligence Review Board (DARB), established in 2022, serves to "operationalize NATO's Principles of Responsible Use of AI" while developing "certification standards" that can ensure compatibility across different national systems. The board's work has become increasingly urgent as the pace of AI adoption accelerates across Allied militaries.

National Approaches Analysis

Canada's Middle Power Strategy: Balancing Innovation and Interoperability

Canada's approach to military AI reflects the classic challenges and opportunities facing middle powers in the contemporary strategic environment. Released in March 2024, Canada's first-ever Artificial Intelligence Strategy for the Department of National Defence and Canadian Armed Forces explicitly acknowledges the need to remain a "credible force that is interoperable with its allies" while developing meaningful AI capabilities.

The strategy emerged from a recognition that Canada's previous approach to AI integration had been "fragmented" and uncoordinated across different military services and defense organizations. As Deputy Minister Bill Matthews and Chief of the Defence Staff General Wayne Eyre noted in their joint statement, "the successful adopting of AI will only come through the transformation and optimization of our data and information management as well as our operational and business practices."

Canada's strategy is built around five key areas designed to transform the Canadian Armed Forces into an "AI-enabled" organization by 2030. These include: identifying and fielding required AI capabilities; establishing ethical frameworks for AI use; managing talent and training needs; deepening strategic partnerships; and ensuring responsible AI governance. This comprehensive approach reflects Canada's recognition that successful AI integration requires simultaneous advances across multiple domains.

The Royal Canadian Navy has emerged as a leader in Canadian military AI experimentation, having requested Defense Research and Development Canada Centre "more than six years ago to set up an AI system that could predict machinery failures on warships." While the system's performance was "not perfect, with false positives also produced," the initial results were "promising enough to warrant further testing," establishing a foundation for broader AI adoption.

Canada's emphasis on alliance interoperability reflects both strategic necessity and opportunity. As a middle power, Canada lacks the resources to develop comprehensive AI capabilities independently. Instead, the strategy explicitly focuses on "leveraging partnerships to access capabilities beyond national resources" while ensuring Canadian forces can contribute meaningfully to collective operations.

The establishment of the Artificial Intelligence Centre in July 2024 represents Canada's most significant organizational innovation in military AI. Designed to "improve alignment with allies, accelerate the adoption of AI capabilities across the defence enterprise, and implement coherent and comprehensive AI solutions," the Centre embodies Canada's approach of treating AI integration as both a technical and strategic challenge.

Canada's participation in key international AI initiatives demonstrates this alliance-focused approach. The Department of National Defence is "a committed member of the AI Partnership for Defence established by the US Department of Defense's Joint Artificial Intelligence Centre," while also maintaining involvement in NATO's evolving AI frameworks and emerging Indo-Pacific partnerships.

Ukraine's Wartime Innovation Model: Necessity as the Mother of Invention

Ukraine's experience since 2022 represents perhaps the most dramatic transformation of military AI capabilities under wartime conditions in modern history. What began as "grassroots applications from 2014-2022" has evolved into a comprehensive AI-enabled military ecosystem that has fundamentally altered the character of contemporary warfare.

The scale of Ukraine's AI-driven military transformation is staggering. In 2024 alone, Ukrainian forces "began purchasing 10,000 AI-enhanced drones" while domestic companies "manufactured and assembled more than 1.5 million FPV drones." Overall, Ukraine produced "approximately 2 million drones in 2024," with "96.2 percent of all UAVs that the Ukrainian armed forces used in 2024" being "produced domestically."

This production surge reflects more than quantitative expansion—it represents a qualitative transformation in how military AI capabilities are developed and deployed. Ukrainian companies have "introduced into their arsenal more than 200 domestically developed unmanned aerial systems and more than 40 ground robotic platforms since February 2022," demonstrating an innovation pace that traditional defense procurement systems struggle to match.

The tactical applications of AI in Ukraine have progressed from basic autonomous navigation to sophisticated target recognition and engagement systems. Ukrainian developers have created drones capable of "locking onto targets identified by operators during the final approach phase," utilizing artificial intelligence "not only at the final targeting stage, but throughout their flight." This evolution has been driven by the practical necessity of countering Russian electronic warfare systems that attempt to disrupt traditional operator-controlled drones.

Ukraine's organizational innovations have proven as significant as its technological advances. The establishment of the national Unmanned Systems Forces in February 2024, with Colonel Vadym Sukharevskyi appointed as commander, represents a formal recognition that AI-enabled autonomous systems require dedicated military structures rather than being treated as adjuncts to traditional forces.

The December 2024 operation near Lyptsi represents a watershed moment in military AI development. Ukrainian forces conducted "the first fully unmanned operation" using "dozens of uncrewed ground vehicles (UGVs) and first-person-view (FPV) drones, with no infantry participation." UGVs equipped with machine guns and munitions performed tasks including "mine clearance and direct fire" while FPV drones provided aerial support, creating a "coordinated multidomain assault" that successfully "destroyed Russian positions."

Ukrainian forces have demonstrated remarkable adaptability in response to enemy countermeasures. As Russian electronic warfare capabilities have evolved, Ukrainian developers have shifted toward "fiber optic drones that are not susceptible to jamming technologies" and AI systems that can "search for, reidentify, and engage targets without a signal or continuous data connection."

The rapid adaptation cycles characterizing Ukrainian AI development have compressed traditional innovation timelines. Where previously Russian forces might take "month-long cycles" to adapt to Ukrainian innovations, they now adjust "in two or three days," creating an accelerated technological competition that demands continuous innovation rather than periodic upgrades.

Ukraine's experience has demonstrated that effective military AI integration requires more than technological capability—it demands organizational agility, cultural adaptation, and the ability to rapidly incorporate lessons learned from combat operations. As Ukrainian officials have stated, "in 2025, more autonomous drones with AI targeting will arrive on the battlefield, potentially making way for real drone swarm uses."

Other Notable Cases: Diverse Paths to AI Integration

Estonia's Digital Defense Model

Estonia represents perhaps the most comprehensive example of how small nations can leverage digital infrastructure for strategic advantage. Building on its experience as one of the world's most digitized societies, Estonia has developed an AI strategy that treats cybersecurity and digital resilience as foundational to national defense.

Estonia's approach reflects lessons learned from the 2007 cyberattacks that paralyzed much of the country's digital infrastructure. The experience served as "a wake-up call for Estonian cyber defense," leading to the creation of the multinational Cooperative Cyber Defence Centre of Excellence and Estonia's emergence as a NATO leader in cyber defense capabilities.

The country's 2024 defense strategy explicitly commits to major investments in military AI, with officials recognizing that "substantive firepower needs to be supported by fast and accurate targeting, alongside timely and adequate decision-making." Estonia's defense spending increase from 3.3 percent to 5 percent of GDP provides resources for "building a digital infrastructure" capable of running "AI solutions" across intelligence analysis, targeting, logistics, and administration.

Estonia's digital defense model benefits from the country's unique position as host to a NATO DIANA "regional hub," second in size only to the agency's London headquarters. This positioning allows Estonia to influence NATO's AI development while ensuring that Estonian innovations are integrated into broader Alliance capabilities.

The country has established a "cyber militia" that coordinates cybersecurity efforts between government and private sector, demonstrating how small nations can mobilize civilian expertise for defense applications. Estonia's State Information Authority has expanded its National Cyber Security Centre (NCSC-EE) while launching initiatives like "Cyber4Dev," an EU-funded cybersecurity program that has "provided assistance to 26 countries."

Singapore's Integrated Approach

Singapore's approach to military AI integration reflects the city-state's broader strategy of leveraging technological sophistication to overcome limitations in size and resources. The country's National Artificial Intelligence Strategy 2.0, launched in 2023, provides the foundation for defense applications by establishing Singapore as "a world leader in AI."

The Singapore Armed Forces (SAF) has undertaken a comprehensive transformation through "SAF 2040," which envisions fully AI-enabled military capabilities supported by the newly established Digital and Intelligence Service (DIS)—a fourth service branch equal in status to the Army, Navy, and Air Force. This organizational innovation demonstrates Singapore's commitment to treating AI as a fundamental military capability rather than a supporting technology.

Singapore's defense AI development benefits from close cooperation with the civilian AI ecosystem. The signing of a Memorandum of Understanding between the DIS and AI Singapore (AISG) creates pathways for "leveraging cutting-edge AI innovations" while building "a robust and resilient pipeline of AI talents that have knowledge of issues related to national defense."

The country has positioned itself as a key player in international AI governance through hosting the "inaugural REAIM Regional Consultations (Asia) in partnership with the Netherlands and the Republic of Korea" and endorsing both "the REAIM Call to Action and the US-led Political Declaration on Responsible Military Use of AI and Autonomy." This positioning allows Singapore to influence global AI norms while advancing its own capabilities.

Singapore's participation in the "AI Partnership for Defense" initiated by the US Department of Defense demonstrates how small nations can leverage great power relationships for capability development. The 2024 signing of a "statement of intent regarding data, analytics and artificial intelligence cooperation" with the United States creates opportunities for Singapore to access advanced AI technologies while contributing specialized capabilities.

UAE's Strategic AI Investment

The United Arab Emirates has pursued an ambitious national AI strategy targeting "regional technological leadership" through sustained investment in education, infrastructure, and capability development. The UAE's approach demonstrates how resource-rich smaller nations can accelerate AI development through strategic partnerships and focused investment.

The country's AI strategy emphasizes specific domain leadership rather than attempting to compete across all areas of AI development. This specialization approach allows the UAE to develop genuine expertise in chosen areas while building relationships with global AI leaders through targeted partnerships and investments.

Israel's Innovation Ecosystem

Israel's military AI development benefits from one of the world's most sophisticated defense technology ecosystems, with seamless integration between military requirements and commercial innovation. The country has developed "advanced AI military applications and autonomous systems" that serve as both national capabilities and export products for the global defense market.

Israel's approach demonstrates how nations with strong technology sectors can leverage military-commercial integration for AI development. The country's experience with AI-based decision support systems, including reported use of systems like "Lavender" in Gaza operations, has generated significant attention and debate regarding the ethics and effectiveness of AI-enabled military decision-making.

Strategic Culture Patterns

Specialization Over Scale: Finding Competitive Niches

The most successful middle power AI strategies share a common recognition: attempting to compete across all domains of AI development leads to resource dispersion and strategic failure. Instead, effective middle powers identify specific niches where they can develop genuine expertise and competitive advantage.

Canada's focus on predictive maintenance and logistics AI reflects this principle. Rather than attempting to develop comprehensive autonomous weapons systems, Canadian forces have concentrated on AI applications that leverage the country's expertise in harsh environment operations and complex logistics challenges. The Royal Canadian Navy's work on predictive maintenance systems for warships represents exactly this type of specialized capability development.

Estonia's concentration on cyber defense and digital resilience exemplifies how small nations can become globally significant through domain specialization. By leveraging its unique experience with large-scale cyberattacks and its position as one of the world's most digitized societies, Estonia has developed AI capabilities in cybersecurity that are valued by much larger allies.

Ukraine's wartime innovation in autonomous systems represents perhaps the most dramatic example of specialization driven by strategic necessity. Faced with a larger, better-equipped adversary, Ukraine has concentrated on developing AI-enabled drones and autonomous systems that maximize the effectiveness of limited human resources while imposing significant costs on enemy forces.

This specialization approach extends beyond technical capabilities to include governance and standards development. The EU's emphasis on regulatory leadership through the AI Act represents a form of specialization—rather than competing in pure technological development, Europe seeks to establish the rules by which global AI systems operate.

Alliance Integration: Leveraging Collective Capabilities

Successful middle powers treat alliance integration not as a constraint on AI development but as a force multiplier that enables access to capabilities beyond national resources. This approach requires careful balancing between maintaining national technological sovereignty and ensuring interoperability with allied systems.

NATO's revised AI strategy explicitly recognizes this challenge, establishing frameworks for "autonomous systems programmed to operate together" while allowing individual nations to maintain distinct technological approaches. The Alliance's emphasis on "growing interoperability between all AI systems throughout the Alliance" creates opportunities for middle powers to contribute specialized capabilities while benefiting from collective development efforts.

Canada's participation in the "AI Partnership for Defence" demonstrates how middle powers can leverage great power relationships for capability development. By aligning Canadian AI development with US Department of Defense initiatives, Canada gains access to advanced technologies and development resources while contributing specialized capabilities in areas like Arctic operations and peacekeeping.

The establishment of DIANA's network across 28 Allied countries creates unprecedented opportunities for technology sharing and joint development. Middle powers can contribute to collective capability development while ensuring their national innovations influence broader Alliance approaches. Estonia's role as host to a DIANA regional hub exemplifies how smaller nations can punch above their weight through strategic positioning in alliance innovation networks.

Singapore's balanced approach to great power relationships—maintaining defense AI cooperation with both the United States and China—demonstrates how middle powers can navigate strategic competition while advancing their own capabilities. This approach requires sophisticated diplomatic management but can provide access to the best technologies from multiple sources.

Regulatory Innovation: Shaping Global Norms

Middle powers have discovered that leadership in AI governance and standards development can provide strategic influence that compensates for limitations in raw technological capability. By establishing ethical frameworks and technical standards, these nations can shape how AI systems are developed and deployed globally.

The EU's AI Act represents the most ambitious example of this approach. By establishing comprehensive regulations for AI development and deployment, Europe seeks to influence global AI development even when European companies may not lead in pure technological capability. This regulatory approach reflects recognition that "whoever sets the AI rules exports their worldview."

NATO's development of AI certification standards through DIANA's network creates opportunities for middle powers to influence how military AI systems are tested and validated. Nations that contribute to these standards development efforts can ensure their approaches and priorities are reflected in Alliance-wide frameworks.

Singapore's hosting of REAIM regional consultations and participation in multiple AI governance initiatives demonstrates how small nations can leverage convening power to influence global AI norms. By positioning itself as a "trusted and substantive interlocutor" in AI governance discussions, Singapore gains influence that extends far beyond its technological capabilities.

France's leadership in promoting "responsible AI in the military domain" through international declarations demonstrates how middle powers can use normative leadership to shape global approaches to military AI. The February 2024 Paris AI Action Summit, which brought together "more than 100 countries, governments, researchers, entrepreneurs and representatives of civil society," exemplifies this approach.

Civilian-Military Fusion: Maximizing Limited Resources

Middle powers have proven particularly adept at breaking down traditional barriers between civilian and military AI development, creating innovation ecosystems that maximize limited resources through dual-use development approaches.

Ukraine's AI ecosystem demonstrates this principle under extreme conditions. The country's rapid military AI development has been enabled by "volunteer groups and private sector" participation, with "more than 200 Ukrainian companies involved in drone production" working directly with military units to "tweak and augment drones to improve their ability to kill and spy on the enemy."

Estonia's "cyber militia" model creates formal pathways for civilian cybersecurity experts to contribute to national defense while maintaining their commercial activities. This approach allows Estonia to mobilize far greater expertise than its military budget could support through traditional procurement approaches.

Canada's emphasis on "deepening strategic partnerships internally and externally with Allies, industries and academia" reflects recognition that effective AI integration requires mobilizing expertise across society rather than relying solely on government resources. The establishment of the Artificial Intelligence Centre represents an attempt to create formal pathways for civilian-military collaboration.

Singapore's integration of defense AI development with its broader national AI strategy ensures that military capabilities benefit from civilian innovation while contributing to broader economic development goals. The DIS's partnership with AI Singapore creates pathways for bidirectional technology transfer between defense and civilian applications.

Case Study Deep Dives

France's AI Military Initiatives: Balancing Sovereignty and Alliance Cooperation

France's approach to military AI represents one of the most comprehensive attempts by a middle power to develop indigenous capabilities while maintaining alliance integration. The country's strategy reflects distinctively French priorities: technological sovereignty, ethical leadership, and strategic autonomy within the broader Western alliance framework.

France's commitment to military AI development is substantial and growing. The military programming law for 2024-2030 "earmarks around €10 billion to accelerate defense innovation with ten technological priorities, including AI and autonomous systems." This represents a significant increase from the "more than €700 million committed over the previous period 2019-2025," demonstrating France's recognition that AI development requires sustained, long-term investment.

The establishment of the Ministerial Agency for Defence AI (AMIAD) in March 2024 represents France's most significant organizational innovation in military AI governance. Funded by "around €300 million per year," AMIAD will coordinate the entirety of France's defense AI development, ensuring coherent strategy implementation across the Ministry of Armed Forces.

France's defense AI strategy, first articulated in the 2019 document "AI in service of defense," established three fundamental principles that continue to guide French development: "abiding by international law, maintaining sufficient human control and ensuring the permanent responsibility of the chain of command." These principles reflect France's attempt to balance military effectiveness with democratic values and international legal obligations.

The Man-Machine Teaming (MMT) study, launched in 2018 and led by Dassault and Thales, exemplifies France's approach to AI integration. The study focuses on "applying AI to combat aircraft" and explores "the feasibility of flying fighter jets and drones together to evade air defense systems." This research reflects France's emphasis on maintaining human-machine teams rather than pursuing fully autonomous systems.

French investment in AI research follows a structured approach, with "roughly half the French annual AI budget" funding studies while "€10 million a year earmarked to test and integrate existing AI technology." Current applications focus on "use cases considered currently more mature, such as detection, recognition, and identification," while longer-term research explores more advanced autonomous capabilities.

France has also emerged as a leader in international military AI governance. The February 2024 Paris AI Action Summit brought together representatives from more than 100 countries to discuss "concrete solutions and initiatives to put AI at the service of the general interest." France's promotion of a declaration on "maintaining human control over AI-enabled weapons systems," supported by 27 states, demonstrates its commitment to responsible AI development.

The French approach faces significant challenges in balancing national sovereignty with alliance integration. Most defense AI programs are "primarily oriented towards hardware development," leaving "little room for increments once they have been launched." This approach complicates the integration of rapidly evolving civilian AI innovations and may limit France's ability to adapt to changing technological landscapes.

Germany's Ethical AI Framework: Deliberative Democracy Meets Military Necessity

Germany's approach to military AI reflects the country's broader strategic culture, which prioritizes democratic deliberation and ethical constraints over rapid capability development. This approach creates both opportunities and constraints for German defense AI development.

Germany's defense AI path has been characterized as caught in a "master and servant logic," where the emphasis on "socio-political acceptance and legitimization of military power" takes precedence over rapid capability development. This reflects Germany's post-war strategic culture, which treats military force as a tool that must be carefully constrained and continuously justified to democratic publics.

Current German spending on defense AI is modest compared to other major European powers. Based on available information, "Germany currently spends around €50 million per year on AI-related software development," significantly less than France's €100 million annual commitment. This reflects both resource constraints and the German government's cautious approach to military AI development.

German defense AI applications currently focus on defensive and support functions rather than offensive capabilities. Publicly known examples include "an open-source intelligence system for crisis early warning, AI-based warning receivers for helicopters, and intelligent image processing for missiles." This emphasis on defensive applications reflects German preferences for technologies that enhance protection rather than offensive capabilities.

The German approach to AI ethics emphasizes "Elements of an Ethical AI Demonstrator for Responsibly Designing Defence Systems," focusing on ensuring that AI development adheres to strict ethical guidelines. German researchers have developed frameworks for "responsible engineering practices" that ensure AI systems remain explainable, reliable, and bias-free.

Germany's military education system is adapting to incorporate AI elements. "The Bundeswehr's Command and Staff College is about to review its curriculum with the goal to incorporate AI elements as of 2024," while "the University of the Bundeswehr/Hamburg is setting up a new AI bachelor's and master's degree course." This educational approach reflects German emphasis on building human understanding of AI rather than simply deploying AI systems.

The German approach faces challenges in balancing ethical constraints with military effectiveness. The country's emphasis on democratic deliberation and public acceptance can slow development timelines, potentially leaving German forces at a disadvantage compared to adversaries with more streamlined decision-making processes.

Germany's position within NATO creates additional complexity. While German ethical frameworks may influence Alliance standards, Germany must also ensure that its systems can interoperate with those of allies who may have different ethical constraints or development timelines.

India's Multi-Alignment Strategy: Navigating Great Power Competition

India's approach to military AI reflects the country's broader strategic doctrine of "multi-alignment," seeking to develop indigenous capabilities while maintaining relationships with multiple global powers. This approach allows India to access the best technologies from various sources while maintaining strategic autonomy.

India's AI development benefits from one of the world's largest technology sectors, with significant expertise in software development, data analytics, and emerging AI technologies. The country's approach seeks to leverage this civilian expertise for defense applications while ensuring that military AI development contributes to broader economic development goals.

India's participation in multiple international AI initiatives demonstrates its multi-alignment approach. The country participates in Quad cooperation on AI standards and development while maintaining separate relationships with Russia, France, and other partners on defense technology cooperation. This approach allows India to access diverse technological approaches while maintaining flexibility in strategic partnerships.

The Indian military has begun integrating AI across multiple domains, from border surveillance systems using computer vision to predictive maintenance applications for complex military equipment. India's focus on AI applications for border security reflects immediate strategic priorities while building broader AI capabilities.

India's approach to AI development emphasizes building indigenous capabilities rather than simply purchasing foreign systems. This reflects broader Indian strategic priorities around "Atmanirbhar Bharat" (self-reliant India) while recognizing the need for international cooperation in rapidly evolving technological domains.

The Indian defense establishment has established research partnerships with leading Indian institutes of technology and civilian AI companies, creating pathways for technology transfer from civilian innovation to military applications. This approach maximizes the military benefits of India's broader AI ecosystem while ensuring that defense applications contribute to civilian technological development.

Innovation Ecosystems: Creating Environments for AI Development

The Ukrainian Model: Wartime Innovation Under Pressure

Ukraine's innovation ecosystem represents perhaps the most dramatic example of how extreme circumstances can accelerate technological development and adoption. The transformation from a primarily agricultural economy to a global leader in military AI development within three years demonstrates the potential for rapid innovation when survival is at stake.

The foundation of Ukrainian success lies in the integration of "volunteer groups and private sector" expertise with military requirements. This integration has created a "24/7 technology race" where "every product in every category must be changed daily to gain an advantage." Such rapid iteration cycles would be impossible within traditional defense procurement systems but have proven essential for maintaining technological superiority against a larger adversary.

Ukrainian Deputy Prime Minister Mykhailo Fedorov's leadership of the "Army of Drones" program demonstrates the importance of high-level political support for innovation. By treating AI development as a national priority equivalent to traditional military operations, Ukraine has been able to mobilize resources and expertise that would otherwise remain in civilian applications.

The establishment of organizations like the Center for Innovation and Defense Technologies (CIDT) within the Ministry of Defense creates formal pathways for integrating civilian innovation with military requirements. The CIDT's responsibility for "upgrading and developing situational awareness technologies, including the Delta system" with integrated AI/ML capabilities demonstrates how institutional innovation can accelerate technology adoption.

Ukraine's approach to regulatory flexibility has been crucial to its success. The Ministry of Digital Transformation has stated that it "does not intend to propose any regulation of AI systems within the defense sector," allowing rapid experimentation and deployment while the war continues. This approach prioritizes capability development over regulatory compliance, though it may create challenges for post-war integration with allied systems.

The Estonian Model: Whole-of-Society Cyber Defense

Estonia's approach to AI innovation reflects the country's unique experience as both a cyber warfare target and a leader in digital government. The integration of civilian cybersecurity expertise with military requirements has created an innovation ecosystem that maximizes limited resources while building capabilities that are valued by much larger allies.

The foundation of Estonian success is the formal integration of civilian and military cybersecurity through the "Cyber Defence Unit under the Estonian Defence League." This volunteer-based organization creates pathways for civilian cybersecurity experts to contribute to national defense while maintaining their commercial activities.

Estonia's position as host to NATO's Cooperative Cyber Defence Centre of Excellence and a DIANA regional hub creates opportunities for the country to influence alliance-wide innovation while ensuring that Estonian innovations are integrated into broader NATO capabilities. This positioning demonstrates how small nations can leverage institutional hosting to punch above their weight in technology development.

The country's investment in AI education and training through initiatives like the "Cybertest platform" and comprehensive cybersecurity training programs creates a foundation for sustained innovation. "Over 15,000 people participated in cybersecurity training programs in 2023," building a workforce capable of supporting ongoing innovation.

Estonia's approach to public-private partnership through initiatives like "CyberTransformation grants" to help small and medium-sized enterprises improve cybersecurity demonstrates how government support can catalyze broader ecosystem development. This approach creates spillover effects that benefit both military and civilian applications.

The Singaporean Model: Strategic Positioning and Excellence

Singapore's innovation ecosystem reflects the city-state's broader strategy of achieving influence through excellence in chosen domains rather than attempting comprehensive coverage across all areas. The country's approach demonstrates how strategic positioning and focused investment can create capabilities that are valued by much larger partners.

The integration of Singapore's civilian AI strategy (NAIS 2.0) with defense requirements through the Digital and Intelligence Service creates pathways for technology transfer while ensuring that defense innovations contribute to broader economic development. This approach maximizes the return on limited resources while building capabilities that serve multiple national objectives.

Singapore's emphasis on international partnerships—including cooperation agreements with both the United States and China—demonstrates how small nations can navigate great power competition while advancing their own capabilities. The July 2024 statement of intent with the United States on "data, analytics and artificial intelligence cooperation" creates opportunities for Singapore to access advanced technologies while contributing specialized capabilities in maritime domain awareness and urban operations.

The country's hosting of international AI governance initiatives, including REAIM regional consultations, positions Singapore as a trusted intermediary in global AI discussions. This positioning creates opportunities for Singapore to influence global AI development while ensuring its own approaches are reflected in international standards.

Singapore's investment in AI education through partnerships between the Digital and Intelligence Service and AI Singapore creates a pipeline of talent with both civilian and military AI expertise. This approach ensures that Singapore can maintain its technological edge while contributing to regional security capabilities.

Key Insight Development: The Middle Power Advantage

The analysis of middle power AI strategies reveals a fundamental insight: successful middle powers are not attempting to become smaller versions of great powers but are instead developing distinctly different approaches that leverage their unique positions in the international system.

Agility Over Scale

Middle powers have demonstrated remarkable agility in AI development, adapting quickly to changing circumstances and technological opportunities. Ukraine's transformation from agricultural economy to military AI leader within three years, Estonia's rapid pivot from cyber warfare victim to NATO cyber defense leader, and Canada's recognition and correction of "fragmented" AI approaches all demonstrate this agility advantage.

This agility stems partly from smaller bureaucratic structures that enable faster decision-making, but more importantly from strategic cultures that embrace adaptation over rigid adherence to established procedures. Middle powers cannot afford the luxury of lengthy deliberation processes when technological change is accelerating and strategic circumstances are shifting rapidly.

Specialization Over Comprehensiveness

Rather than attempting to compete across all domains of AI development, successful middle powers identify specific niches where they can develop genuine expertise and competitive advantage. This specialization approach allows limited resources to achieve maximum impact while creating capabilities that are valued by larger allies.

Estonia's focus on cyber defense, Ukraine's concentration on autonomous systems, Canada's emphasis on Arctic operations and predictive maintenance, and Singapore's development of urban and maritime AI applications all demonstrate how specialization can create genuine competitive advantages that transcend resource limitations.

Integration Over Independence

Middle powers have recognized that AI development in the contemporary strategic environment requires integration with broader international networks rather than pursuit of technological independence. This integration takes multiple forms: alliance cooperation through NATO and bilateral partnerships, civilian-military fusion within national innovation ecosystems, and public-private collaboration that mobilizes expertise across society.

The most successful middle powers treat this integration not as a constraint on sovereignty but as a multiplier that enables access to capabilities and resources beyond what any single nation could achieve independently. NATO's DIANA network, bilateral AI cooperation agreements, and civilian-military innovation partnerships all demonstrate how integration can enhance rather than constrain national capabilities.

Governance Over Pure Technology

Middle powers have discovered that leadership in AI governance and standards development can provide strategic influence that compensates for limitations in raw technological capability. The EU's AI Act, NATO's responsible use principles, and various international AI governance initiatives demonstrate how middle powers can shape the rules by which AI systems operate globally.

This governance leadership reflects a recognition that AI development is not purely a technical challenge but a socio-political one that requires addressing questions of ethics, accountability, and democratic oversight. Middle powers' experience with democratic governance and alliance cooperation positions them well to provide leadership in these areas.

The strategic cultures emerging among middle powers in the AI age share common characteristics: emphasis on adaptation over rigidity, specialization over comprehensiveness, integration over independence, and governance leadership over pure technological competition. These approaches reflect not limitations but strategic choices that leverage middle powers' unique positions in the international system.

As AI continues to reshape the strategic landscape, middle powers that embrace these approaches are likely to achieve influence and effectiveness that exceeds their material capabilities. The challenge for larger powers will be recognizing and adapting to these new forms of strategic competition, while the opportunity for middle powers lies in continuing to innovate in approaches to AI development and integration.

The future strategic landscape will likely be characterized not by the dominance of a few AI superpowers but by a complex ecosystem where middle powers play crucial roles through specialization, innovation, and leadership in governance frameworks. Understanding and adapting to this reality will be essential for all nations seeking to navigate the AI-transformed strategic environment successfully.

Sources and References:

- NATO. "NATO releases revised AI strategy." July 10, 2024. https://www.nato.int/cps/en/natohq/news_227234.htm
- NATO. "Summary of NATO's revised Artificial Intelligence (AI) strategy." July 10, 2024. https://www.nato.int/cps/en/natohq/official_texts_227237.htm
- NATO. "Defence Innovation Accelerator for the North Atlantic (DIANA)." https://www.nato.int/cps/en/natohq/topics_216199.htm
- Canadian Armed Forces. "The Department of National Defence and Canadian Armed Forces Artificial Intelligence Strategy." March 2024. https://www.canada.ca/en/department-national-defence/corporate/reports-publications/dnd-caf-artificial-intelligence-strategy.html
- CBC News. "New DND strategy warns Canadian military's approach to artificial intelligence 'fragmented'." July 29, 2024. https://www.cbc.ca/news/politics/military-artificial-intelligence-strategy-1.7277628
- Centre for International Governance Innovation. "Militarizing AI: How to Catch the Digital Dragon?" 2024. https://www.cigionline.org/articles/militarizing-ai-how-to-catch-the-digital-dragon/
- Digital Watch Observatory. "Cybersecurity Strategy 2024–2030: Cyber-Conscious Estonia." February 24, 2025. https://dig.watch/resource/cybersecurity-strategy-2024-2030-cyber-conscious-estonia
- Breaking Defense. "Estonia pledges major investments in military AI." April 24, 2025. https://breakingdefense.com/2025/03/estonia-pledges-major-investments-in-military-ai/
- CSIS. "Ukraine's Future Vision and Current Capabilities for Waging AI-Enabled Autonomous Warfare." March 20, 2025. https://www.csis.org/analysis/ukraines-future-vision-and-current-capabilities-waging-ai-enabled-autonomous-warfare
- Atlantic Council. "Missiles, AI, and drone swarms: Ukraine's 2025 defense tech priorities." January 2, 2025. https://www.atlanticcouncil.org/blogs/ukrainealert/missiles-ai-and-drone-swarms-ukraines-2025-defense-tech-priorities/
- Modern War Institute. "Battlefield Drones and the Accelerating Autonomous Arms Race in Ukraine." January 10, 2025. https://mwi.westpoint.edu/battlefield-drones-and-the-accelerating-autonomous-arms-race-in-ukraine/

- CSIS. "Understanding the Military AI Ecosystem of Ukraine." January 21, 2025. https://www.csis.org/analysis/understanding-military-ai-ecosystem-ukraine
- U.S. Trade.gov. "Singapore Artificial Intelligence Strategy 2.0." March 18, 2024. https://www.trade.gov/market-intelligence/singapore-artificial-intelligence-strategy-20
- Breaking Defense. "US-Singapore boost artificial intelligence ties, after small nation made earlier agreement with China." July 17, 2024. https://breakingdefense.com/2024/07/us-singapore-boost-artificial-intelligence-ties-after-small-nation-made-earlier-agreement-with-china/
- TechWire Asia. "Singapore military to enhance digital defenses with AI." May 15, 2023. https://techwireasia.com/2023/05/singapore-military-to-enhance-digital-defenses-with-ai/
- SpringerLink. "A Winding Road Before Scaling-Up? Defense AI in France." 2024. https://link.springer.com/chapter/10.1007/978-3-031-58649-1_11
- Research Professional News. "France announces R&D agency for military artificial intelligence." March 21, 2024. https://www.researchprofessionalnews.com/rr-news-europe-france-2024-3-france-announces-r-d-agency-for-military-ai/
- France UN. "France is mobilized to build an inclusive international governance of AI." 2024. https://onu.delegfrance.org/france-is-mobilized-to-build-inclusive-and-sustainable-international-governance
- SpringerLink. "Master and Servant: Defense AI in Germany." 2024. https://link.springer.com/chapter/10.1007/978-3-031-58649-1_9
- American German Institute. "Germany Needs to Consider Military AI." September 2020. https://americangerman.institute/2020/09/germany-needs-to-consider-military-ai/
- European Parliament. "Defence and artificial intelligence." 2025. https://www.europarl.europa.eu/thinktank/en/document/EPRS_BRI(2025)769580

Chapter 8: Small States, Smart Strategies: AI as the Great Leveler

How AI Democratizes Strategic Capabilities

Opening Scene: The Great Equalizer in Action

On June 1, 2025, at precisely 02:00 local time across five Russian oblasts spanning five time zones, Ukrainian Security Service (SBU) operatives executed what would become known as Operation Spider's Web—the most audacious demonstration to date of how artificial intelligence has fundamentally democratized strategic military capabilities.

The operation was breathtaking in its scope and sophistication. Ukrainian drones, smuggled deep into Russian territory and concealed within commercial trucks parked near military airfields, were remotely activated from command centers hundreds of kilometers away in Ukraine. Using AI-enhanced guidance systems and real-time target recognition algorithms, these relatively inexpensive commercial drones—each costing mere hundreds of dollars—struck at the heart of Russia's strategic nuclear deterrent.

The results were devastating for Moscow. According to Ukrainian officials, over 40 Russian military aircraft were destroyed or damaged across five airbases, including Tu-160 Blackjack strategic bombers, Tu-95 Bear long-range bombers, and Tu-22M3 Backfire intermediate-range bombers. The Financial Times reported that the damaged and destroyed aircraft represented approximately 20% of Russia's operational long-range aviation fleet. Most significantly, these aircraft—many dating from the Soviet era and no longer in production—were effectively irreplaceable.

What made this operation historically significant was not merely its tactical success, but what it demonstrated about the changing nature of warfare in the AI age. As one Western defense analyst observed, "A small country with ingenuity and relatively low-cost drones had just delivered a strategic blow to a nuclear superpower that conventional wisdom said was impossible." The operation, which took 18 months to plan and execute, proved that with artificial intelligence enhancement, small states could achieve strategic effects previously available only to great powers.

The implications reverberated through defense establishments worldwide. As noted in *The Washington Post*, "The audacious attacks on Russian airfields highlight the power of unconventional, asymmetric warfare—and the threat to U.S. strategic bases and ports." The operation demonstrated that AI had become the great leveler in international relations, enabling small states to punch far above their weight and challenge traditional assumptions about power projection and strategic deterrence.

This chapter examines how artificial intelligence is democratizing strategic capabilities, enabling small states to develop asymmetric advantages that can offset conventional military disadvantages.

Through detailed case studies of Estonia's digital defense model, Singapore's integrated AI approach, the UAE's strategic AI investments, and Israel's innovation ecosystem, we explore how nations with limited resources are leveraging AI to achieve disproportionate strategic impact.

The Democratization Effect: When David Gets Smart

The transformation of warfare through artificial intelligence represents one of the most significant democratization events in military history since the advent of gunpowder. Traditional military advantages—large armies, expensive platforms, extensive industrial bases—are being challenged by the AI-enabled capabilities that small, innovative states can develop and deploy with remarkable speed and efficiency.

Breaking the Great Power Monopoly

For decades, advanced military capabilities were the exclusive domain of major powers with extensive defense-industrial complexes. Developing precision-guided munitions, sophisticated surveillance systems, or strategic strike capabilities required massive investments in research, development, and manufacturing infrastructure that only wealthy nations could afford. This technological barrier effectively limited the strategic options available to smaller states, forcing them into defensive postures or dependence on great power patrons.

Artificial intelligence has fundamentally altered this equation. As Ukrainian forces demonstrated throughout 2024 and 2025, commercial AI technologies can now provide military-grade capabilities that rival or exceed traditional defense systems. The democratization effect operates through several key mechanisms:

Commercial AI Providing Military-Grade Capabilities: Modern commercial AI systems, particularly in computer vision, autonomous navigation, and target recognition, have reached levels of sophistication that were classified military technologies just a decade ago. Ukrainian drone manufacturers have proven that off-the-shelf AI components can be integrated into military systems with devastating effectiveness. As reported by the Atlantic Council, Ukrainian forces moved beyond the initial concept of "one drone, one operator" to deploy swarm technologies that can operate autonomously throughout their flight paths.

Reduced Barriers to Advanced Capabilities: The cost and complexity barriers that traditionally limited access to advanced military technologies have been dramatically reduced. AI-powered drones that cost hundreds of dollars can now accomplish missions that previously required million-dollar missile systems. The proliferation of open-source AI tools, commercial sensor technologies, and 3D printing capabilities has enabled rapid prototyping and deployment of sophisticated military systems by relatively small organizations.

User-Friendly Interfaces Masking Complexity: Perhaps most significantly, modern AI systems provide intuitive interfaces that mask underlying technological complexity. Ukrainian operators have demonstrated the ability to conduct precision strikes against high-value targets using interfaces described as similar to commercial video games, while the AI handles complex navigation, target recognition, and terminal guidance functions autonomously.

The Ukraine Model: Proof of Concept

Ukraine's military transformation since 2022 provides the most compelling evidence of AI's democratizing effect on warfare. Beginning with grassroots initiatives that included volunteer groups and private sector innovation, Ukraine rapidly scaled AI-enabled capabilities that have fundamentally altered the strategic balance in its conflict with Russia.

The numbers are striking. According to Bloomberg reporting from June 2025, "by early 2025, drones were accounting for 60% to 70% of the damage and destruction caused to Russian equipment in the war." This represents a complete transformation of warfare, where relatively inexpensive AI-enabled systems have become the primary instruments of tactical and strategic effect.

Ukraine's domestic drone production capabilities exemplify this transformation. The country aims to produce approximately four million drones in 2025, more than double the previous year's total. The Ukrainian Ministry of Defense allocated over $2.5 billion in 2024-2025 for local drone manufacturers, signing contracts with 76 companies. This represents a remarkable scaling of capability from a standing start in 2022.

The strategic implications extend beyond mere tactical success. As noted by Ukrainian officials, the goal is to "remove warfighters from direct combat" through autonomous systems. This represents a fundamental shift from human-intensive to AI-intensive warfare, one that small states can implement more rapidly than larger, more bureaucratic military establishments.

Technological Convergence Enabling Small State Innovation

Several technological trends have converged to enable this democratization effect:

Artificial Intelligence Integration: The integration of AI into commercial electronics has reached a point where sophisticated machine learning capabilities are available in consumer devices. This has enabled small state militaries to integrate advanced AI into weapons systems without requiring dedicated AI research programs.

Sensor Miniaturization: The smartphone revolution has driven massive improvements in the miniaturization and cost reduction of sensors, cameras, and communication systems. These components, originally developed for consumer markets, provide military-grade sensing capabilities at commercial prices.

Communication Technologies: Satellite internet systems like Starlink have provided small state forces with resilient, high-bandwidth communication capabilities that were previously available only to major powers with dedicated military satellite constellations.

Manufacturing Democratization: 3D printing and other advanced manufacturing technologies have enabled rapid prototyping and small-batch production of military systems, reducing the advantage of traditional defense-industrial complexes.

The convergence of these technologies has created an environment where small states can rapidly develop and deploy sophisticated military capabilities using primarily commercial technology foundations, enhanced with AI to achieve military-grade performance.

Case Studies in Small State Innovation

Estonia's Digital Defense Model: Leading Through Cyber Resilience

Estonia represents perhaps the most successful example of a small state leveraging digital technologies and artificial intelligence to achieve strategic influence far exceeding its geographic and demographic limitations. With a population of just 1.3 million, Estonia has positioned itself as a global leader in cybersecurity and digital governance, providing a model for other small states seeking to maximize their strategic impact through technological innovation.

Foundation of Digital Resilience

Estonia's strategic approach to AI and cybersecurity is built on a foundation of comprehensive digitalization that began in the 1990s. The country's experience of cyber attacks, particularly the 2007 attacks that targeted government, banking, and media websites, galvanized national commitment to developing robust cyber defense capabilities. Rather than viewing this challenge as insurmountable, Estonia leveraged the crisis to build world-leading expertise in cyber resilience.

The Estonian model demonstrates how small states can turn vulnerability into strategic advantage. As Gert Auväärt, Deputy Director-General for Cybersecurity at the Estonian Information System Authority (RIA), noted in the country's 2025 cybersecurity report, "Estonia faced an unprecedented number of distributed denial-of-service attacks, but effective defence measures reduced the proportion of attacks with an impact."

AI Integration Across Government and Defense

Estonia's approach to military AI integration exemplifies how small states can leverage civilian technological capabilities for defense applications. In March 2025, the Estonian Ministry of Defense announced a comprehensive AI strategy calling for building "digital infrastructure" robust enough to run "AI solutions" for everything from intelligence analysis and wartime targeting to smart logistics and back-office administration.

The strategy represents a holistic approach to AI integration that leverages Estonia's digital government infrastructure. As Ambassador Kristjan Prikk explained at the 2025 AITalks conference, "Whatever we lack in size, we can compensate for by using data smarter." Estonia is developing a new cross-government, AI-enabled data management tool designed to expand citizens' access to federal information and resources, with military applications integrated into this broader digital infrastructure.

Estonia's defense AI spending reflects serious commitment to technological transformation. The country announced it would increase overall defense spending from 3.3 percent of GDP to 5 percent, with significant portions dedicated to AI capabilities. This represents one of the highest defense spending ratios among NATO allies, demonstrating how small states can achieve strategic relevance through focused investment in emerging technologies.

International Leadership and Alliance Integration

Estonia has successfully translated its AI and cybersecurity expertise into international influence. The country hosts NATO's Cooperative Cyber Defence Centre of Excellence, positioning it as a thought leader in alliance cyber defense strategy. According to the Global Cybersecurity Index, Estonia ranks first in the EU on cybersecurity capabilities, providing it with soft power influence far exceeding its size.

The country's partnership with German defense AI startup Helsing, announced in August 2024, demonstrates how small states can access cutting-edge technologies through strategic partnerships. The agreement includes a €70 million investment from Helsing to establish operational facilities in Estonia, focusing on "software-defined reconnaissance and strike capabilities." This collaboration exemplifies how small states can become attractive partners for international technology companies by offering regulatory environments and testing opportunities that larger bureaucracies cannot match.

Estonia's role in the NATO DIANA (Defence Innovation Accelerator for the North Atlantic) network further demonstrates its strategic positioning. Tallinn hosts a "regional hub" for the accelerator, second in size only to London headquarters, providing Estonia with influence over alliance technology development priorities.

Lessons from the Estonian Model

Estonia's success provides several key lessons for other small states seeking to leverage AI for strategic advantage:

Early Investment in Digital Infrastructure: Estonia's current AI capabilities are built on decades of investment in digital government and civilian technology infrastructure. This foundation enables rapid scaling of AI applications across government and defense sectors.

Public-Private Integration: Estonia's "cyber militia" concept, which coordinates private sector expertise with government cybersecurity needs, provides a model for civilian-military AI integration that maximizes limited national resources.

International Partnership Strategy: Rather than attempting to develop all capabilities domestically, Estonia strategically partners with international technology companies and alliance partners to access advanced capabilities while maintaining sovereignty over key applications.

Regulatory Innovation: Estonia's development of AI governance frameworks positions it as a standard-setter for responsible AI development, providing soft power influence that enhances its strategic relevance.

Singapore's Integrated Approach: Smart Nation as Defense Foundation

Singapore represents a unique model of how small states can integrate artificial intelligence across civilian and military applications to achieve comprehensive national transformation. The city-state's Smart Nation initiative, launched in 2014 and expanded with Smart Nation 2.0 in 2024, demonstrates how systematic AI integration can enhance both economic competitiveness and national security capabilities.

Strategic Vision and Implementation

Singapore's approach to AI reflects a characteristically comprehensive and long-term strategic vision. The country's National AI Strategy 2.0, launched in December 2023, aims to triple AI talent from 5,000 to 15,000 professionals and establish Singapore as a global leader in AI development and deployment by 2030. This civilian AI development provides the foundation for military applications through what Singaporean officials describe as "dual-use technology development."

The scope of Singapore's AI ambitions is reflected in its investment commitments. The Smart Nation 2.0 initiative includes a S$120 million fund specifically for AI adoption, while the broader research and development ecosystem has attracted over 150 AI research and development teams from leading global companies and research institutions.

Singapore's approach demonstrates how small states can leverage their unique advantages—in this case, regulatory agility and geographic position—to become regional technology hubs. As noted in academic analysis of Singapore's AI ecosystem, "by designing policies that address risks associated with AI development and implementation, smaller countries can create a vibrant AI ecosystem that encourages experimentation and early adoption of the technology."

Military AI Applications and Defense Integration

Singapore's defense establishment has systematically integrated AI capabilities across multiple domains, reflecting the country's recognition that technological superiority is essential for maintaining security in a complex geopolitical environment. The Singapore Armed Forces (SAF) established an AI laboratory under the Defence Science and Technology Agency in 2017, focusing on developing artificial intelligence and analytics for defense applications.

In 2021, Singapore unveiled a new command and control system that uses AI and data analytics to recommend weapons and support faster, more effective decision-making processes. This system represents the kind of integrated approach that leverages civilian AI research for military applications while maintaining appropriate security boundaries.

The Defense Science and Technology Agency (DSTA) plays a crucial role in this integration, facilitating cutting-edge research on AI, robotics, unmanned vehicles, cybersecurity, and advanced materials while collaborating with academia and industry partners. This organizational approach enables Singapore to leverage its limited resources efficiently by avoiding duplication between civilian and military AI research efforts.

Governance Framework and International Positioning

Singapore has positioned itself as a leader in AI governance, developing frameworks that influence global standards while serving its national interests. The Model AI Governance Framework for Generative AI (MGF-GenAI), developed in collaboration with international partners, seeks to create comprehensive and trusted AI ecosystems that address global risks associated with generative AI.

This regulatory leadership provides Singapore with strategic influence in international forums. The country's participation in the Global Partnership on AI (GPAI) and its alignment with various international AI governance initiatives demonstrate how small states can shape global norms through technological expertise and diplomatic engagement.

Singapore's approach to AI governance reflects its broader strategic philosophy of maintaining positive relationships with all major powers while preserving sovereignty and strategic autonomy. As noted in research on Singapore's AI strategy, the country "has adopted a highly pragmatic approach in dealing with the great AI rivalry between the United States and China," leveraging partnerships with both superpowers while maintaining strategic independence.

Economic-Security Integration

Singapore's Smart Nation initiative demonstrates how small states can integrate economic development and security objectives through AI development. The Punggol Digital District, described by Prime Minister Lawrence Wong as "Singapore's first-ever smart district," brings together companies, research institutes, and government agencies working in cybersecurity, AI, and robotics.

This integrated approach enables Singapore to maximize the strategic value of its AI investments by ensuring that civilian technological developments support both economic competitiveness and defense capabilities. The co-location of research, development, and application activities creates innovation ecosystems that can rapidly translate technological advances into operational capabilities.

The economic foundation of Singapore's AI strategy is robust, with government data indicating that the country is home to over 150 AI R&D and product teams from leading global companies and research institutions. The government's "AI for Science" program encourages interdisciplinary collaborations between AI researchers and scientists in other domains, creating the kind of cross-pollination that enables breakthrough applications.

UAE's Strategic AI Investment: Regional Leadership Through Technology

The United Arab Emirates represents an ambitious model of how resource-rich small states can leverage financial capabilities to achieve technological leadership and strategic influence. The UAE's National Strategy for Artificial Intelligence 2031 demonstrates how focused investment and strategic partnerships can position small states as regional technology leaders while enhancing defense capabilities.

Vision and Strategic Objectives

The UAE's AI strategy is characterized by remarkable ambition and comprehensive scope. The National AI Strategy 2031 aims to position the UAE as a global leader in AI by 2031, with the goal of generating AED 335 billion (approximately $91 billion) in economic growth through AI integration across healthcare, education, transportation, and defense sectors.

The strategy's eight strategic objectives demonstrate the comprehensive nature of the UAE's approach: building reputation as an AI destination, increasing competitive assets through AI deployment, developing a fertile AI ecosystem, adopting AI across government services, attracting and training AI talent, bringing world-leading research capabilities to target industries, ensuring strong governance and effective regulation, and establishing the UAE as a world leader in AI applications.

The UAE made history in 2017 as the first country worldwide to create a ministry dedicated to artificial intelligence, demonstrating early recognition of AI's strategic importance. The creation of the AI and Advanced Technology Council in January 2024 further strengthened institutional commitment to AI development across government and defense sectors.

Defense and Security Applications

The UAE's approach to defense AI reflects its broader strategic philosophy of leveraging technological superiority to enhance security in a complex regional environment. The country has invested significantly in cybersecurity infrastructure and defense technology, with defense spending reaching $23.2 billion in 2023 and projected cumulative defense investments of $129 billion between 2024 and 2028.

The UAE's defense AI development is characterized by strategic partnerships with international technology companies and defense contractors. Recent collaborations include partnerships with companies developing autonomous systems, cyber defense capabilities, and AI-enhanced surveillance systems. These partnerships enable the UAE to access cutting-edge technologies while building domestic capabilities.

The UAE's investment in defense technology innovation is demonstrated by its participation in international defense exhibitions and conferences. At IDEX 2025, the UAE showcased advanced AI-enabled systems including autonomous ground vehicles, AI-enhanced surveillance platforms, and integrated command and control systems that demonstrate the country's commitment to technological military modernization.

Energy Infrastructure and AI Capabilities

The UAE's strategic approach to AI is underpinned by substantial investment in energy infrastructure that provides the foundation for large-scale AI operations. According to analysis from PwC Middle East, the UAE's future electricity generation capacity will significantly outstrip domestic demand, creating opportunities for energy-intensive AI applications including defense systems.

By 2030, the UAE is expected to have 26.6 gigawatts of total generation capacity while domestically consuming only 21.2 gigawatts. This surplus energy capacity, generated from a mix of renewable, nuclear, and fossil fuel sources, provides the foundation for extensive AI operations including military applications that require substantial computational resources.

The UAE's Barakah nuclear power plant, when fully operational, will provide additional energy capacity that could support advanced AI applications. If the UAE constructs a second nuclear facility with similar specifications by 2035, total generation capacity could reach 34.4 gigawatts, providing enormous capacity for AI-intensive operations.

International Partnerships and Strategic Positioning

The UAE's AI strategy is characterized by strategic partnerships with global technology leaders that provide access to cutting-edge capabilities while supporting domestic development objectives. Microsoft's $1.5 billion investment in Abu Dhabi's G42 demonstrates growing international confidence in the UAE's AI ecosystem and strategic positioning.

In September 2024, MGX, together with U.S. firms BlackRock, Global Infrastructure Partners, and Microsoft, launched a global AI infrastructure fund with $30 billion in initial private equity capital and $100 billion in investment potential. More recently, in January 2025, OpenAI announced that MGX would be one of the initial equity funders for the "Stargate Project," which plans to invest $500 billion in building U.S. AI infrastructure.

These partnerships demonstrate how small states can leverage financial resources to gain access to cutting-edge technologies and influence global technology development. The UAE's strategy of becoming an essential partner for major technology companies provides it with access to advanced capabilities while supporting its domestic AI development objectives.

Israel's Innovation Ecosystem: Military-Civilian AI Integration

Israel represents perhaps the most sophisticated example of how small states can develop advanced AI capabilities through systematic integration of military, academic, and private sector innovation. With a population of approximately 9.5 million, Israel has achieved global leadership in AI-enabled defense technologies through what experts describe as a "dual feeding" process that links military service, academic research, and commercial innovation.

Institutional Framework and Organization

Israel's approach to military AI is characterized by sophisticated institutional structures that facilitate rapid development and deployment of advanced technologies. In January 2025, the Israeli Ministry of Defense announced the creation of a new research unit within the Directorate of Defense Research and Development (Mafat) focused specifically on artificial intelligence and autonomy.

According to Reserve Major General Eyal Zamir, Director General of the Ministry of Defense, this initiative represents "a pivotal development in the preparation of future battlefields, where soldiers and autonomous systems will operate side by side." The new AI and Autonomy

Administration aims to centralize expertise and foster collaboration across military branches, academic institutions, startups, and major defense companies.

The sophistication of Israel's approach is demonstrated by the integration of AI across multiple military domains. As noted by defense officials, AI-powered algorithms analyze radar and sensor data to track incoming missiles and calculate optimal interception timing in systems like Iron Dome and David's Sling. These systems have achieved success rates exceeding 90 percent while reducing operating costs, demonstrating the practical effectiveness of AI integration.

Military-Academic-Commercial Integration

Israel's success in military AI reflects unique cultural and institutional factors that facilitate rapid innovation and technology transfer. Elite military technology units, particularly Unit 8200 (signals intelligence), recruit talented high school graduates who receive extensive training in advanced technologies. Upon completion of military service, many of these individuals join startup companies or establish their own firms, often in cybersecurity and AI applications.

This "dual feeding" process creates powerful synergies between military needs and commercial innovation. According to data presented at defense technology conferences, since October 2023, the Israeli army has collaborated with 86 Israeli startups and small companies on contracts worth approximately $168 million—five times more than during the same period the previous year.

The commercial applications of military AI research extend far beyond defense. Companies founded by veterans of elite military units have achieved global success in cybersecurity, autonomous systems, and AI applications. Notable examples include Unit 81 graduates who have established approximately 50 companies between 2003 and 2010, including firms like Innoviz and D-Fend that provide solutions used for both civilian and defense applications.

Operational AI Applications

Israel's military AI applications demonstrate the practical effectiveness of systematic technology integration. The Israeli Defense Forces acknowledge using AI-based tools for targeting support, intelligence analysis, proactive forecasting, and streamlined command and control. These systems have been deployed in operational contexts including the 2021 Guardian of the Walls operation and subsequent conflicts.

The "Gospel" (Habsora) system, which generates possible military targets for attack, represents one of the most sophisticated AI targeting systems deployed in operational contexts. According to military officials, AI-based systems successfully identified commanders of Hamas's anti-aircraft and missile units from substantial pools of potentially threatening individuals during recent operations.

Border control applications demonstrate the comprehensive scope of Israeli military AI deployment. The IDF uses AI systems to assist border observers through video analysis that identifies individuals, vehicles, animals, and specific threats including armed individuals and particular vehicle models. These systems incorporate facial recognition capabilities and real-time threat assessment algorithms.

Global Influence and Technology Export

Israel's military AI capabilities have significant global influence through technology exports and strategic partnerships. Israel is one of the world's largest weapons exporters, with AI-enhanced systems being deployed by military forces worldwide. Israeli companies have exported lethal UAVs to Chile, China, India, South Korea, and Turkey, spreading Israeli AI military technologies globally.

The global influence of Israeli military AI extends beyond direct exports. Israeli spyware and surveillance technologies have been adopted by various governments worldwide, demonstrating how small state technological innovations can achieve global reach. This technological influence provides Israel with soft power capabilities that enhance its strategic position despite geographic and demographic limitations.

Israel's investment in developing lethal autonomous weapon systems (LAWS) positions it at the forefront of next-generation military technologies. The country's Harpy loitering munition is globally recognized as one of the most advanced autonomous weapons systems, providing a model for other small states seeking to develop similar capabilities.

Lessons from Ukraine's AI Evolution: The Grassroots to Government Pipeline

Ukraine's transformation from a conventional military force to an AI-enabled asymmetric power provides the most comprehensive real-world case study of how small states can rapidly develop and deploy advanced military AI capabilities under pressure. The Ukrainian experience offers crucial lessons for other small states seeking to leverage AI for strategic advantage.

Foundation Building: Grassroots Innovation (2014-2022)

Ukraine's current AI capabilities were not developed overnight but built on a foundation of grassroots innovation that began during the 2014 conflict in Donbas. Volunteer groups and private sector organizations began experimenting with commercial drones and basic AI applications for military purposes, creating the technological and cultural foundation for later scaling.

This grassroots foundation proved crucial when full-scale conflict began in 2022. As noted by Ukrainian officials, "volunteer groups and private sector" involvement in AI development during the preparatory period enabled rapid integration and scaling when military necessity demanded it. The informal nature of early development allowed for experimentation and learning that would have been difficult in traditional military bureaucracies.

The volunteer-driven development model created cultural changes within Ukrainian military forces that facilitated AI adoption. Soldiers became comfortable with technological innovation and adaptation, creating organizational cultures that could rapidly integrate new AI capabilities as they became available.

Wartime Acceleration and Scaling

The period from 2022 to 2025 demonstrates how wartime pressure can accelerate AI development and deployment at unprecedented rates. Ukraine's military AI expansion across six major applications since 2022 represents one of the most rapid technology integration programs in military history.

The numbers reflect this dramatic scaling. Ukraine aims to produce approximately four million drones in 2025, supported by UAH 56 billion (USD 1.5 billion) allocated for defense industry development. This represents a transformation from essentially zero domestic military AI capability to mass production in less than three years.

The Ukrainian experience demonstrates how operational necessity can overcome bureaucratic obstacles to technology integration. Traditional military procurement processes, which typically require years of testing and evaluation, were compressed into months or weeks as battlefield necessity demanded rapid adaptation.

Rapid Adaptation Cycles and Organizational Learning

Perhaps the most significant lesson from Ukraine's experience is the importance of rapid adaptation cycles in AI military applications. Ukrainian innovations have forced Russian forces to adapt to new technologies and tactics in increasingly short timeframes. According to Ukrainian military sources, Russian adaptation cycles have compressed from month-long periods to "two or three days" as the pace of technological change accelerates.

This acceleration of adaptation cycles represents a fundamental change in military dynamics. Traditional military planning, which assumed relatively stable technological environments, must be replaced with continuous adaptation processes that can integrate new AI capabilities as they emerge.

The Ukrainian model demonstrates the importance of organizational learning in AI military applications. Ukrainian forces have systematically captured lessons from AI deployments and rapidly integrated improvements into new systems. This learning process is facilitated by close cooperation between military users and technology developers, enabling rapid feedback and iteration.

Technology Transfer and Scaling

Ukraine's experience provides insights into how AI military technologies can be transferred and scaled across different contexts. Ukrainian innovations in drone swarm technologies, autonomous targeting systems, and electronic warfare countermeasures have been studied and adapted by military forces worldwide.

The Ukrainian approach to technology development—emphasizing adaptability, cost-effectiveness, and rapid deployment—provides a model for other small states seeking to develop AI military capabilities. Rather than attempting to match the capabilities of major powers directly, Ukrainian forces have focused on asymmetric applications that leverage AI to achieve disproportionate effects.

The integration of commercial technologies with military applications represents another key lesson from Ukrainian experience. Ukrainian forces have demonstrated how consumer electronics, commercial software, and civilian AI research can be rapidly adapted for military purposes, reducing development costs and timeframes.

International Support and Technology Transfer

Ukraine's AI development has been significantly enhanced by international support and technology transfer. The twenty-nation Drone Coalition co-chaired by Latvia and the UK has pledged €2.75 billion to supply an additional one million drones in 2025, demonstrating how international cooperation can accelerate small state AI capabilities.

The Ukrainian experience demonstrates how small states can leverage international partnerships to access advanced technologies and capabilities that would be difficult to develop independently. Strategic partnerships with technology companies, academic institutions, and allied governments have enabled Ukraine to achieve AI capabilities that would normally require much larger resource investments.

Strategic Implications: Redefining Power in the AI Age

The success of small states in developing and deploying advanced AI military capabilities has profound implications for international relations and strategic stability. Traditional assumptions about the relationship between state size, resources, and military capability are being challenged by the democratizing effects of artificial intelligence.

From Survival to Specialization

Historically, small states focused primarily on survival strategies—defensive postures designed to deter aggression and maintain sovereignty. The AI revolution has enabled a shift from survival to specialization, where small states can develop particular technological advantages that provide strategic influence beyond their geographic and demographic limitations.

Estonia's leadership in cybersecurity, Singapore's integrated approach to AI governance, the UAE's energy-intensive AI capabilities, and Israel's military-civilian innovation ecosystem all demonstrate how small states can achieve strategic relevance through specialized technological capabilities rather than comprehensive military power.

This shift from survival to specialization represents a fundamental change in small state strategic options. Rather than accepting dependency on great power patrons, small states can develop autonomous capabilities that provide strategic influence and bargaining power in international relations.

Asymmetric Capabilities and Strategic Effects

AI-enabled systems have demonstrated the ability to achieve strategic effects that are disproportionate to their cost and complexity. Ukraine's Operation Spider's Web exemplifies this dynamic—relatively inexpensive drones achieved strategic effects against Russian nuclear capabilities that would normally require major military operations.

The asymmetric nature of AI capabilities creates new strategic dynamics where small states can threaten or deter much larger adversaries. Traditional metrics of military power—numbers of tanks, aircraft, or soldiers—become less relevant when small, AI-enabled systems can achieve decisive effects against high-value targets.

This asymmetric capability development has implications for alliance relationships and strategic stability. Small states with advanced AI capabilities become more valuable allies, while the proliferation of these capabilities may create new forms of strategic instability as traditional deterrence relationships are challenged.

Enhanced Alliance Value and Strategic Partnerships

Small states with advanced AI capabilities are becoming increasingly valuable alliance partners. Estonia's role in NATO cybersecurity, Singapore's position as a regional technology hub, and Israel's technology exports demonstrate how AI capabilities enhance small state strategic value to larger partners.

The enhanced alliance value creates new forms of strategic leverage for small states. Rather than being passive recipients of security guarantees, small states with advanced AI capabilities can provide essential technologies and expertise that major powers require for their own strategic objectives.

This dynamic creates opportunities for small states to shape alliance policies and strategic priorities through technological leadership rather than traditional military contributions. The influence is particularly significant in areas where small states have achieved technological advantages over larger partners.

Regional Technology Leadership

Several small states have achieved regional technology leadership through focused AI development, creating new forms of influence and strategic positioning. The UAE's AI infrastructure investments, Singapore's Smart Nation initiative, and Estonia's digital government model all demonstrate how technological leadership can translate into regional influence.

Regional technology leadership provides small states with soft power capabilities that can influence neighbor countries' policies and strategic choices. Countries that successfully position themselves as technology leaders can attract investment, talent, and partnerships that enhance their overall strategic position.

The competitive dynamics of regional technology leadership create incentives for continued innovation and investment, potentially accelerating AI development across entire regions as countries compete for technological advantage.

New Strategic Cultures Emerging

The successful integration of AI capabilities by small states is generating new forms of strategic culture that differ significantly from traditional military organizations. These emerging strategic cultures emphasize rapid adaptation, civilian-military integration, and technological innovation over conventional military virtues.

Rapid Innovation and Adaptation Cultures

Small states leading in AI development have cultivated organizational cultures that prioritize rapid innovation and adaptation over traditional military virtues of stability and predictability. Estonian cybersecurity organizations, Ukrainian military-technology integration, and Israeli defense innovation all demonstrate cultures that embrace experimentation and rapid change.

These cultures are characterized by flattened hierarchies that enable rapid decision-making, tolerance for failure as part of learning processes, and systematic integration of lessons learned into operational procedures. Traditional military cultures, which emphasize standardization and predictability, are being supplemented or replaced by cultures that value adaptability and innovation.

The emergence of rapid innovation cultures has implications for military education, professional development, and organizational structure. Military institutions must develop capabilities for continuous learning and adaptation rather than mastery of stable doctrines and procedures.

Civilian Innovation Ecosystem Integration

Successful small states have systematically broken down barriers between military and civilian innovation ecosystems, creating integrated approaches that leverage all available technological capabilities. Israel's military-academic-commercial integration, Singapore's Smart Nation approach, and Estonia's public-private partnerships all demonstrate this integration.

This integration requires cultural changes in both military and civilian organizations. Military forces must become comfortable with civilian technologies and development processes, while civilian organizations must understand military requirements and security constraints.

The integration of civilian innovation ecosystems provides small states with access to global technology developments and commercial innovation resources that would be unavailable through purely military research and development programs.

Domain-Specific Excellence

Rather than attempting to develop comprehensive military capabilities, successful small states have focused on achieving excellence in specific technological domains. This approach reflects both resource constraints and strategic recognition that specialized capabilities can provide disproportionate strategic influence.

Estonian cybersecurity expertise, Israeli autonomous systems development, Singapore's AI governance frameworks, and Ukrainian asymmetric warfare innovations all demonstrate how domain-specific excellence can provide strategic advantages that transcend traditional measures of military power.

The focus on domain-specific excellence requires cultural changes that prioritize depth over breadth in capability development. Military organizations must develop tolerance for dependencies in some areas while pursuing leadership in others.

Continuous Learning and Experimentation

Emerging strategic cultures in AI-leading small states emphasize continuous learning and experimentation over adherence to established doctrines. Ukrainian military adaptation, Estonian cyber defense evolution, and Israeli technology integration all demonstrate cultures that systematically learn from experience and integrate improvements into operational procedures.

This emphasis on continuous learning requires organizational structures and processes that can capture lessons, analyze effectiveness, and implement improvements rapidly. Traditional military cultures, which often emphasize adherence to established procedures, must develop capabilities for systematic innovation and adaptation.

The culture of continuous learning and experimentation provides small states with adaptive advantages that can enable them to maintain technological edges despite resource constraints.

Future Trajectory Analysis: The Strategic Landscape in 2030-2040

The trajectory of small state AI development suggests that the democratizing effects of artificial intelligence will continue to accelerate through the next decade, potentially fundamentally altering the international strategic landscape.

Technological Convergence and Capability Proliferation

Current trends suggest that AI capabilities will continue to proliferate and improve rapidly, making advanced military technologies accessible to an increasing number of small states. The convergence of AI with other emerging technologies—quantum computing, advanced materials, biotechnology—will create new opportunities for small state innovation and capability development.

The pace of technological change is likely to favor small states that can adapt quickly over larger bureaucracies that require extensive coordination and approval processes. Small states with agile decision-making processes and experimental cultures may be able to integrate emerging technologies faster than traditional military powers.

The proliferation of AI capabilities will likely create a more multipolar strategic environment where numerous small states possess capabilities that can influence regional and global security dynamics.

Regional AI Blocs and Technology Alliances

The development of regional AI capabilities may lead to the formation of technology alliances and blocs that create new forms of strategic alignment. Countries with complementary AI capabilities may form partnerships that enhance collective technological capabilities while reducing individual development costs.

Regional AI blocs may compete for technological leadership, access to resources, and influence over global AI development standards. This competition could accelerate innovation while creating new forms of strategic rivalry that supplement traditional geopolitical competition.

The formation of technology alliances may provide small states with collective capabilities that enable them to compete with major powers in specific technological domains.

Democratization Limits and Emerging Constraints

While AI has democratized many military capabilities, certain constraints may limit the extent of this democratization. Access to advanced semiconductors, rare earth materials, and energy resources may create new forms of technological dependence that limit small state autonomy.

The development of AI governance frameworks and export controls by major powers may attempt to limit the proliferation of advanced AI capabilities to small states. The effectiveness of these measures will depend on the ability of major powers to control global technology supply chains and development processes.

Some AI applications may require resources or capabilities that remain available only to major powers, creating limits to the democratizing effects of artificial intelligence.

Strategic Stability and Deterrence Evolution

The proliferation of AI capabilities among small states will likely require evolution of deterrence theories and strategic stability concepts. Traditional deterrence relationships based on nuclear weapons and conventional force balances may be supplemented by considerations of AI capabilities and cyber vulnerabilities.

Small states with advanced AI capabilities may develop new forms of deterrent threats based on their ability to disrupt major power systems or achieve asymmetric effects. These capabilities may enhance strategic stability by increasing the costs of aggression, or may decrease stability by creating new vulnerabilities and escalation pathways.

The evolution of strategic stability concepts will require international cooperation and potentially new arms control frameworks that address AI military applications while preserving the legitimate security interests of small states.

Challenges and Limitations

Despite the significant opportunities that AI provides for small states, several challenges and limitations constrain the extent to which these capabilities can enhance strategic influence and security.

Resource Constraints and Sustainability

While AI capabilities can be developed more affordably than traditional military systems, they still require significant investments in research, development, and deployment. Small states must balance AI investments against other national priorities, creating constraints on the scale and scope of capability development.

The sustainability of AI capabilities requires ongoing investments in maintenance, upgrading, and adaptation. Small states may find it difficult to maintain technological edges over time as competitors develop countermeasures and alternative approaches.

Resource constraints may force small states to make difficult choices between different AI applications, limiting their ability to develop comprehensive capabilities across multiple domains.

Dependency Risks and Technology Access

Small states developing AI capabilities may become dependent on international suppliers for critical components, software, and expertise. These dependencies create vulnerabilities that can be exploited by major powers through export controls, sanctions, or other forms of economic pressure.

Access to cutting-edge AI technologies may be controlled by major powers seeking to limit the proliferation of advanced capabilities. Small states may find themselves excluded from access to the most advanced AI systems and development tools.

The global nature of AI supply chains creates multiple points of potential vulnerability where small state access to essential technologies can be disrupted by geopolitical tensions or economic competition.

Talent and Expertise Limitations

AI development requires specialized human capital that may be difficult for small states to attract and retain. Competition for AI talent from major powers and commercial companies may create brain drain effects that limit small state capabilities.

The interdisciplinary nature of AI development requires expertise across multiple domains—computer science, engineering, domain-specific knowledge—that may be difficult for small countries to maintain across all relevant areas.

Educational and training infrastructure requirements for AI development may strain small state resources and require international partnerships that create dependencies.

Technological Obsolescence and Adaptation Pressure

The rapid pace of AI development creates constant pressure for technological adaptation and upgrading. Small states may struggle to maintain currency with rapidly evolving technologies while managing other national priorities.

The competitive nature of AI development means that technological advantages can be rapidly eroded as competitors develop countermeasures or alternative approaches. Small states must maintain continuous innovation to preserve strategic advantages.

The integration of AI capabilities with legacy systems and organizational structures may create technical challenges that are difficult for small states to manage without extensive external support.

Key Insight Development: The Transformation of Strategic Competition

The emergence of small states as significant actors in AI-enabled strategic competition represents one of the most significant transformations in international relations since the end of the Cold

War. The democratizing effects of artificial intelligence are creating new possibilities for small state influence while challenging traditional assumptions about the relationship between state size and strategic capability.

The Ukrainian experience in particular demonstrates that with ingenuity, determination, and appropriate technological choices, small states can achieve strategic effects that were previously available only to major powers. The ability of relatively inexpensive AI-enabled systems to achieve decisive effects against high-value targets fundamentally alters strategic calculations for both small states and their larger neighbors.

However, this transformation is not without risks and limitations. The proliferation of AI capabilities may create new forms of strategic instability as traditional deterrence relationships are challenged. The dependencies created by global technology supply chains may limit small state autonomy even as AI capabilities enhance their strategic options.

The success of small states in AI development depends critically on their ability to develop appropriate strategic cultures that can effectively integrate civilian and military innovation while maintaining rapid adaptation capabilities. Those small states that can create these cultural foundations are likely to achieve strategic influence far exceeding their traditional capabilities.

The international community must develop new frameworks for managing the strategic implications of AI proliferation while preserving the legitimate security interests of small states. This will require evolution of alliance relationships, deterrence concepts, and potentially new forms of arms control that address the unique characteristics of AI military applications.

The transformation of strategic competition through AI democratization is likely to continue accelerating through the next decade, creating both opportunities and challenges for small states seeking to enhance their security and influence in an increasingly complex international environment.

Conclusion: The Great Leveling

The emergence of artificial intelligence as a democratizing force in international relations represents a profound shift in the global strategic landscape. Small states that were once relegated to passive roles in international security are now developing capabilities that enable them to influence regional and global dynamics through technological innovation rather than traditional military power.

The case studies examined in this chapter—from Ukraine's dramatic demonstration of asymmetric AI warfare to Estonia's cyber leadership, Singapore's integrated development approach, the UAE's strategic AI investments, and Israel's innovation ecosystem—demonstrate that success in the AI age depends more on strategic vision, organizational culture, and technological adaptation than on traditional measures of state power.

The lessons learned from these experiences provide roadmaps for other small states seeking to leverage AI for strategic advantage. The emphasis on rapid innovation cultures, civilian-military integration, and specialized technological development offers alternatives to traditional approaches to small state security.

However, the democratization of AI capabilities also creates new challenges and risks. The proliferation of these capabilities may generate new forms of strategic instability, while dependencies on global technology supply chains create vulnerabilities that can be exploited by larger powers.

The future strategic landscape will likely be characterized by increased influence for small states that successfully integrate AI capabilities while larger powers must adapt to operating in environments where technological surprise and asymmetric capabilities play increasingly important roles.

The AI revolution has fundamentally altered the strategic playing field, creating opportunities for small states to punch above their weight while challenging traditional assumptions about power, deterrence, and strategic stability. The nations that recognize and adapt to these changes will shape the international order of the coming decades.

Sources and References

1. **Operation Spider's Web Documentation**
    - Wikipedia. "Operation Spider's Web." https://en.wikipedia.org/wiki/Operation_Spider's_Web
    - NPR. "Ukraine destroys more than 40 military aircraft in a drone attack deep inside Russia." June 1, 2025. https://www.npr.org/2025/06/01/nx-s1-5419509/ukraine-destroys-military-aircraft-attack-inside-russia-planes
    - The Washington Post. "Ukraine's surprise drone strike underscores vulnerability of U.S. defenses." June 4, 2025. https://www.washingtonpost.com/national-security/2025/06/04/ukraine-drone-strike-asymmetric-future-warfare/
2. **Ukraine AI Development**
    - Atlantic Council. "Missiles, AI, and drone swarms: Ukraine's 2025 defense tech priorities." January 2, 2025. https://www.atlanticcouncil.org/blogs/ukrainealert/missiles-ai-and-drone-swarms-ukraines-2025-defense-tech-priorities/
    - Atlantic Council. "A Western-funded drone surge could end Russia's invasion of Ukraine." July 13, 2025. https://www.atlanticcouncil.org/blogs/ukrainealert/a-western-funded-drone-surge-could-end-russias-invasion-of-ukraine/
    - Bloomberg. "How Ukraine's Drone Arsenal Shocked Russia and Changed Modern Warfare." June 3, 2025. https://www.bloomberg.com/features/2025-ukraine-drones-explainer/

3. **Estonia Digital Defense and AI Strategy**
    - Digital Watch Observatory. "Cybersecurity Strategy 2024–2030: Cyber-Conscious Estonia." February 24, 2025. https://dig.watch/resource/cybersecurity-strategy-2024-2030-cyber-conscious-estonia
    - Breaking Defense. "Estonia pledges major investments in military AI." April 24, 2025. https://breakingdefense.com/2025/03/estonia-pledges-major-investments-in-military-ai/
    - e-estonia.com. "Cyber security in the age of AI and hybrid threats." March 19, 2025. https://e-estonia.com/cyber-security-in-the-age-of-ai-and-hybrid-threats/
    - DefenseScoop. "Estonia moves to develop new cross-government, AI-powered data management asset." April 25, 2025. https://defensescoop.com/2025/04/25/estonia-ai-powered-data-management-tool-ambassador-kristjan-prikk/
4. **Singapore Smart Nation and AI Strategy**
    - Smart Nation Singapore. "National Artificial Intelligence Strategy 2 to uplift Singapore's social and economic potential." December 4, 2023. https://www.smartnation.gov.sg/media-hub/press-releases/04122023/
    - PMO Singapore. "PM Lawrence Wong at the Launch of Smart Nation 2.0." October 1, 2024. https://www.pmo.gov.sg/Newsroom/PM-Lawrence-Wong-at-the-Launch-of-Smart-Nation
    - Nature. "Building an AI ecosystem in a small nation: lessons from Singapore's journey to the forefront of AI." July 2024. https://www.nature.com/articles/s41599-024-03289-7
    - SpringerLink. "Reimagining Defense Innovation: Defense AI in Singapore." 2024. https://link.springer.com/chapter/10.1007/978-3-031-58649-1_25
5. **UAE AI Strategy and Investments**
    - Digital Watch Observatory. "The UAE National Strategy for Artificial Intelligence 2031." November 18, 2024. https://dig.watch/resource/the-uae-national-strategy-for-artificial-intelligence-2031
    - Chambers and Partners. "AI in UAE: The Legal Blueprint That's Reshaping Tech Compliance in 2025." 2025. https://chambers.com/articles/ai-in-uae-the-legal-blueprint-thats-reshaping-tech-compliance-in-2025
    - CSIS. "The United Arab Emirates' AI Ambitions." January 27, 2025. https://www.csis.org/analysis/united-arab-emirates-ai-ambitions
    - UAE Government. "UAE Strategy for Artificial Intelligence." https://u.ae/en/about-the-uae/strategies-initiatives-and-awards/strategies-plans-and-visions/government-services-and-digital-transformation/uae-strategy-for-artificial-

intelligence

6. **Israel AI Military Applications**
   - Defense News. "Israel creates hub to hasten military AI, autonomy research." January 2, 2025. https://www.defensenews.com/global/mideast-africa/2025/01/02/israel-creates-hub-to-hasten-military-ai-autonomy-research/
   - Army Recognition. "Israel Strengthens its Innovation Capacity in Artificial Intelligence and Speeds Up Integration of New Technologies." January 2025. https://armyrecognition.com/news/army-news/2025/israel-strengthens-its-innovation-capacity-in-artificial-intelligence-and-speeds-up-integration-of-new-technologies
   - Lieber Institute West Point. "Beyond the Headlines: Combat Deployment of Military AI-Based Systems by the IDF." September 6, 2024. https://lieber.westpoint.edu/beyond-headlines-combat-deployment-military-ai-based-systems-idf/
   - Times of Israel. "AI a 'game-changer' for future of warfare, says Defense Ministry director-general." December 10, 2024. https://www.timesofisrael.com/ai-a-game-changer-for-future-of-warfare-says-defense-ministry-director-general/
7. **Regional and International Analysis**
   - EV Magazine. "Estonia Leads Global Defence Innovation Through Alliances." March 4, 2025. https://evmagazine.com/news/estonia-leads-global-defence-innovation-through-alliances
   - Strategy International. "The Fusion of Technology and Defense: Israel's Military-Technology Complex." November 21, 2024. https://strategyinternational.org/2024/11/21/publication150/
   - Russia Matters. "Russia in Review, May 30-June 6, 2025." June 2025. https://www.russiamatters.org/news/russia-review/russia-review-may-30-june-6-2025

Chapter 9: Nuclear Strategy in the Age of Algorithms

The Ultimate Human Decision Meets Machine Speed

A few terrifying moments separated the world from nuclear war. On the morning of January 24, 2025, in a CSIS conference room in Washington D.C., that historical reality hung heavy in the air as two of America's leading defense experts faced off in a heated debate about whether artificial intelligence should have any role in nuclear command and control.

"We need to acknowledge that there are several different and varying definitions of what artificial intelligence is," began Sarah Mineiro, senior associate for the Aerospace Security Project at CSIS, speaking in favor of AI integration into nuclear command, control, and communications (NC3) systems. "The term 'artificial intelligence' means a machine-based system that can, for a given set of human-defined objectives, make predictions, recommendations, or decisions influencing real or virtual environments."

On the opposite side, Paul Scharre, executive vice president and director of studies at the Center for New American Security, countered with a chilling question that would define the stakes: "What is the training data set we have for nuclear war?" His point was devastating in its simplicity—unlike self-driving cars that have highways full of data to learn from, nuclear AI systems would be making life-or-death decisions based on theoretical scenarios with virtually no real-world examples to guide them.

The debate, hosted by CSIS's Project on Nuclear Issues (PONI), crystallized a fundamental tension at the heart of modern strategic culture: how to balance the need for speed in an age of hypersonic missiles and compressed decision timelines against the ultimate requirement for human wisdom in the most consequential decisions imaginable. As Heather Williams, director of PONI, framed the motion: "The United States should increase its reliance on artificial intelligence to enhance decision-making in its NC3 systems."

The timing of this debate was no coincidence. Just months earlier, U.S. Strategic Command Commander General Anthony Cotton had stated that AI implementation makes U.S. Nuclear Command, Control and Communications "more resilient to adversaries, and strengthens deterrence," while simultaneously emphasizing the need to maintain "human decision in the loop." The 2022 Nuclear Posture Review had already affirmed U.S. policy to "maintain a human 'in the loop' for all actions critical to informing and executing decisions by the President to initiate and terminate nuclear weapon employment."

Yet as Mineiro pointed out in the debate, an intercontinental ballistic missile "takes at most half an hour to reach its target"—and modern threats like hypersonic weapons compress that timeline even further. In such scenarios, she argued, AI could "expand decision space for human decision makers" by rapidly analyzing "petabytes of data" and processing intelligence far faster than human analysts ever could. Every minute spent manually analyzing potential threats, she contended, is "one minute less to make crucial judgment calls."

But Scharre's response cut to the core of the risks: AI systems acting "in unexpected manner providing faulty but seemingly accurate information" to decision-makers who desperately need reliable data in crisis moments. The stakes, as both debaters acknowledged, could not be higher—the difference between accurate threat assessment and algorithmic error could be the difference between peace and nuclear war.

This debate encapsulates the central challenge facing all nuclear-armed nations as they grapple with integrating artificial intelligence into their most sensitive strategic systems. It represents what may be the ultimate test of how strategic cultures adapt to the AI revolution—where the margin for error is quite literally zero, and where the speed of modern warfare increasingly conflicts with the deliberative nature of democratic decision-making.

The Current Debate: A Sharp Division

The nuclear policy community finds itself sharply divided on perhaps the most consequential question of the AI age: should artificial intelligence play any role in the systems that control humanity's most destructive weapons? This division runs deeper than typical policy disagreements—it reflects fundamentally different philosophies about the relationship between human judgment and machine capabilities in ultimate strategic decisions.

The Fundamental Divide

The debate over nuclear AI represents more than a technical disagreement; it reveals a philosophical chasm about the nature of strategic decision-making itself. On one side stand those who argue that the increasing speed of modern warfare—driven by hypersonic weapons, stealth cruise missiles, and compressed decision timelines—necessitates AI assistance to provide human decision-makers with the information they need to make informed choices in crisis scenarios.

General Anthony Cotton, commander of U.S. Strategic Command, represents this pragmatic view. In his October 2024 address at the Department of Defense Intelligence Information System conference, Cotton argued that "advanced AI and robust data analytics capabilities provide decision advantage and improve our deterrence posture." He emphasized that AI would "enhance our decision-making capabilities" while maintaining clear boundaries: "But we must never allow artificial intelligence to make those decisions for us."

This position reflects what Cotton described as a fundamental modernization imperative. As he told audiences at the Center for Strategic and International Studies in November 2024, "If we think that United States Strategic Command can't take advantage of artificial intelligence to preserve the terabytes of data that would otherwise hit the floor, and still do things the old fashioned way...we might as well move out of the beautiful building we have and just go into something that has rotary phones."

On the opposing side stand those who view any AI integration into nuclear systems as an unacceptable risk to human control over ultimate strategic decisions. This camp, led by experts like Paul Scharre, argues that the unique nature of nuclear weapons—their unprecedented destructive power and the irreversible nature of their use—demands that humans retain complete control over all aspects of nuclear decision-making.

The Arms Control Association has articulated this position clearly, warning that AI integration with nuclear command and control "could mitigate human error or bias in nuclear decision-making" but simultaneously "raise the chance of nuclear use in the most dangerous manner possible: subtly and without warning." The concern is that AI systems might provide seemingly accurate but fundamentally flawed assessments that could lead decision-makers toward catastrophic miscalculations.

Official U.S. Policy Framework

Current U.S. policy attempts to thread this needle by maintaining what officials call "meaningful human control" over nuclear decisions while exploring AI applications in supporting roles. The Biden Administration's 2022 Nuclear Posture Review established the foundational principle: "In all cases, the United States will maintain a 'human in the loop' for all actions critical to informing and executing decisions by the President to initiate and terminate nuclear weapon employment."

This policy was reinforced in October 2024 in the Administration's Framework to Advance AI Governance and Risk Management in National Security, which reiterated the human-in-the-loop requirement for nuclear systems. However, the framework leaves significant room for interpretation regarding what constitutes "critical" actions and how AI might assist in non-critical but still important functions.

The distinction matters because, as Sarah Mineiro argued in the CSIS debate, there are numerous functions within NC3 systems that could benefit from AI assistance without crossing the line into autonomous decision-making about weapons use. These include intelligence processing, threat assessment, communications routing, and system diagnostics—all areas where AI could potentially improve speed and accuracy while keeping humans firmly in control of ultimate decisions.

International Perspectives and Commitments

The international community has increasingly rallied around the principle that nuclear weapons should never be subject to fully autonomous control. The United Nations has been particularly vocal on this issue, with Secretary-General António Guterres repeatedly stating that nuclear weapons should "always include humans within decision-making chains."

In November 2024, a significant diplomatic breakthrough occurred when Chinese President Xi Jinping and U.S. President Joe Biden jointly affirmed "the need to maintain human control over the decision to use nuclear weapons." This marked the first time China had made such a public commitment, reflecting growing international consensus that nuclear decisions must remain under human authority.

The joint U.S.-China statement built on earlier declarations, including a 2022 document submitted by France, the United Kingdom, and the United States during the Nuclear Non-Proliferation Treaty review process that emphasized human control over nuclear weapons. These international commitments represent a rare area of agreement among major powers, even as they compete intensely in other domains of AI development.

However, the practical implementation of these commitments remains challenging. As arms control experts note, distinguishing between "decision support" AI that assists human judgment and "decision-making" AI that effectively controls outcomes can be extremely difficult in practice. The concern is that incremental steps toward greater AI integration might gradually erode human control without any single decision crossing a clear red line.

The Training Data Problem

Perhaps the most compelling argument against nuclear AI integration is what Paul Scharre calls "the training data problem." Unlike other military applications where AI systems can learn from extensive real-world data—such as conventional warfare, where there are numerous historical examples to draw from—nuclear conflict presents a unique challenge: there have been only two instances of nuclear weapons use in warfare, both in 1945, and virtually no examples of nuclear crisis management that provide clear templates for AI learning.

This data scarcity creates multiple risks. First, AI systems trained primarily on simulated data may not perform reliably when confronted with real-world nuclear crises that inevitably include unexpected variables and human factors that simulations cannot capture. Second, the lack of historical examples means that AI systems cannot learn from past successes and failures in nuclear decision-making, potentially making them less reliable than human decision-makers who can draw on broader strategic intuition and contextual understanding.

Moreover, as University of Pennsylvania political scientist Michael Horowitz points out, the phenomenon of "automation bias" means that human operators tend to trust automated systems even when those systems provide incorrect information. In a nuclear context, this could lead to what Horowitz calls a scenario where "the Petrov of the future doesn't use his judgment, or that there is no Petrov"—a reference to Soviet Lieutenant Colonel Stanislav Petrov, whose human judgment prevented nuclear war in 1983 when he correctly identified a false alarm from automated systems.

The Speed Versus Deliberation Dilemma

The fundamental tension in nuclear AI debates centers on the conflict between the need for speed in modern warfare and the requirement for careful deliberation in nuclear decisions. Proponents of AI integration argue that emerging technologies—hypersonic weapons, stealth cruise missiles, and space-based threats—are compressing decision timelines to the point where human analysis alone may be insufficient.

As Adam Lowther and Curtis McGiffin argued in a controversial 2019 War on the Rocks article, "Technologies such as hypersonic weapons, stealthy nuclear-armed cruise missiles, and weaponized artificial intelligence mean America's legacy NC3 system may be too slow for the president to make a considered decision and transmit orders." They suggested that AI systems might need to pre-authorize certain responses to cope with compressed timelines.

However, critics argue that this emphasis on speed fundamentally misunderstands the nature of nuclear deterrence. Unlike conventional warfare, where rapid response can provide tactical advantages, nuclear weapons are primarily political instruments whose effectiveness depends on the credibility of threatened retaliation rather than the speed of that retaliation. From this perspective, rushing nuclear decisions to match the pace of incoming threats may be counterproductive if it increases the risk of false alarms or inappropriate escalation.

The speed-versus-deliberation debate also reflects different views of how nuclear crises are likely to unfold. Those favoring AI integration often assume scenarios of surprise attack requiring immediate response, while those opposing it emphasize that most nuclear risks arise from escalating conventional conflicts where there should be time for human deliberation and diplomatic communication.

AI Applications Under Consideration

Despite the heated debates about nuclear AI, military planners and policymakers are already exploring specific applications where artificial intelligence might enhance nuclear command and control without crossing red lines about human decision-making authority. These applications range from intelligence processing and early warning enhancement to decision support systems that could provide commanders with better information while preserving human control over ultimate choices.

Early Warning System Enhancement

The most widely discussed application of AI in nuclear systems involves enhancing early warning capabilities that detect incoming threats. Current systems, based on satellite sensors and ground-based radars, process enormous amounts of data to distinguish between potential threats and false alarms. AI could potentially improve both the speed and accuracy of this threat detection process.

As noted in Arms Control Association analysis, AI's ability to find correlations by "continuously sorting through large amounts of data with an unbiased eye" makes it particularly relevant for early-warning systems. Machine learning systems could "produce significant benefits by ensuring accurate, timely information for decision-makers and by recognizing potential attacks" while helping assess "the increasingly sophisticated data being relayed to the North American Aerospace Defense (NORAD) Command in Colorado."

The potential benefits are substantial. Modern early warning systems must process satellite imagery, radar signatures, and sensor data from multiple sources to build comprehensive threat pictures. AI algorithms could potentially identify patterns and correlations that human analysts might miss, while simultaneously reducing the time required to assess potential threats. This could be particularly valuable given that, as Sarah Mineiro noted in the CSIS debate, every minute spent analyzing threats is "one minute less to make crucial judgment calls."

However, AI enhancement of early warning systems also carries significant risks. As experts at the Arms Control Association warn, AI's technical limitations "could compound decades-old problems of alerting attacks where there is none." The concern is that AI systems might create new categories of false positives while providing them with a veneer of algorithmic authority that makes human operators less likely to question their assessments.

The challenge becomes particularly acute when considering that AI systems are designed to identify patterns in data, but nuclear crises often involve unprecedented situations that don't match historical patterns. An AI system trained to recognize "normal" patterns of military activity might misinterpret routine exercises or unusual but innocent activities as preparation for attack, potentially triggering alerts in situations where human analysts would exercise appropriate skepticism.

Intelligence Processing and Fusion

Beyond early warning, AI applications in nuclear command and control could extend to intelligence processing and fusion—the complex task of integrating information from multiple sources to create comprehensive situational awareness. General Cotton has been particularly vocal about AI's potential in this area, describing how AI could help Strategic Command process "terabytes of data that would otherwise hit the floor."

In his November 2024 CSIS presentation, Cotton explained how AI could be valuable "in increasing efficiency in intelligence, surveillance, and reconnaissance" while emphasizing that "it's not about having a machine present forces to the president of the United States. It's just the opposite of that." Instead, AI would help human analysts process and understand complex intelligence pictures more quickly and accurately.

The potential applications are extensive. AI systems could help correlate satellite imagery with communications intercepts, identify patterns in adversary military movements, and integrate data from multiple intelligence sources to provide commanders with comprehensive threat assessments. In crisis situations, this enhanced intelligence processing could help decision-makers understand adversary intentions and capabilities more quickly than current manual processes allow.

Paul Scharre, despite his overall skepticism about nuclear AI, acknowledged in the CSIS debate that AI could help "crunch numbers and do pattern recognition, classification, and flight path determination" more quickly than human analysts. The key distinction, he argued, is ensuring that

AI assists human judgment rather than replacing it—providing processed information to human decision-makers rather than making recommendations about appropriate responses.

However, intelligence fusion also presents significant risks for AI integration. Intelligence analysis often requires understanding context, cultural factors, and political dynamics that may be difficult for AI systems to grasp. An AI system might correctly identify military movements but misinterpret their significance, leading to threat assessments that are technically accurate but strategically misleading.

Decision Support Systems

Perhaps the most controversial AI application involves decision support systems that would help senior officials evaluate response options in nuclear crises. These systems would go beyond simple intelligence processing to provide analysis of potential courses of action, their likely consequences, and their relationship to broader strategic objectives.

Air Force officials have indicated that they see AI providing "actionable insights and enabling better informed decisions" in nuclear contexts. As General Cotton described it, AI systems could help commanders respond when the president asks for options: "I [would] much rather say, 'well, Mr. President, hold on. I'll get back to you in a couple of hours, and we'll talk about how we can execute that'" rather than providing immediate but potentially inadequate responses.

Decision support AI could potentially model the effects of different nuclear responses, assess the likely reactions of adversaries and allies, and help commanders understand the broader implications of various courses of action. In complex scenarios involving multiple adversaries or alliance considerations, AI systems might help decision-makers navigate the intricate strategic calculations involved in nuclear crisis management.

The appeal of such systems is obvious—nuclear decision-making involves extraordinarily complex calculations involving military effectiveness, political signaling, alliance dynamics, and escalation management. AI systems could potentially process these multiple variables more quickly and comprehensively than human analysts working under extreme time pressure and stress.

However, decision support systems also represent the area where AI integration most directly challenges human control over nuclear decisions. The line between "supporting" human decision-making and "making" decisions can become blurred when AI systems provide specific recommendations about nuclear responses. There's a significant risk that human decision-makers, facing enormous pressure and limited time, might defer to AI recommendations without adequate independent judgment.

Moreover, nuclear decision-making often involves considerations—moral, political, and strategic—that may be difficult to quantify in ways that AI systems can process. The decision to use nuclear weapons involves not just military calculations but judgments about acceptable risk, moral responsibility, and long-term consequences that may require uniquely human forms of reasoning.

Communications and System Management

Less controversial but potentially equally important are AI applications in managing the communications and technical systems that comprise the NC3 infrastructure. These systems must function reliably under all conditions, including during nuclear conflict, making their resilience and redundancy critical to maintaining effective deterrence.

AI could potentially enhance NC3 system management in several ways. Machine learning algorithms could predict system failures before they occur, automatically route communications around damaged or compromised nodes, and optimize network performance under degraded conditions. During conflicts, AI systems could help maintain communications connectivity even as adversaries attempt to disrupt or destroy NC3 infrastructure.

The Pentagon's focus on "cyber resilience" in NC3 systems makes AI applications particularly relevant for defending against cyber attacks. AI systems could potentially detect and respond to cyber intrusions more quickly than human operators, automatically implementing countermeasures to preserve system integrity during conflicts.

However, even these seemingly technical applications raise important questions about human control. If AI systems are automatically making decisions about communications routing or cyber defense measures, they could potentially affect the information available to human decision-makers or the ability of senior officials to communicate with military commanders. Ensuring that such AI systems enhance rather than compromise human control requires careful attention to their design and implementation.

The Implementation Challenge

Regardless of specific applications, implementing AI in nuclear systems faces fundamental challenges that go beyond technical capabilities. The most significant is ensuring that AI integration enhances rather than undermines the reliability and controllability of nuclear systems.

As researchers at War on the Rocks note, successful human-machine teams in nuclear contexts must "harness the precision and speed of automation with the flexibility of human judgment" while avoiding "automation bias and surrendering human judgment to machines." This requires not just technical solutions but organizational and cultural changes to ensure that human operators maintain appropriate skepticism about AI recommendations.

The challenge is compounded by the fact that nuclear systems must function reliably for decades with minimal opportunity for real-world testing. Unlike other military AI applications that can be refined through battlefield experience, nuclear AI systems must work correctly the first time they're needed in a real crisis—a requirement that places extraordinary demands on their design, testing, and validation.

Risks and Concerns Analysis

The integration of artificial intelligence into nuclear command and control systems presents a complex web of risks that could fundamentally alter the nature of nuclear deterrence and crisis stability. These risks span technical vulnerabilities, strategic miscalculations, and systemic failures that could lead to catastrophic consequences. Understanding these dangers is essential for policymakers grappling with the balance between AI's potential benefits and its existential risks in nuclear contexts.

Technical Vulnerabilities and Algorithmic Failures

The most immediate concerns about nuclear AI stem from the inherent limitations and vulnerabilities of current artificial intelligence technologies. These technical risks are particularly acute in nuclear contexts because of the irreversible nature of nuclear decisions and the unprecedented stakes involved.

AI Hallucinations and False Information Generation

Modern AI systems, particularly large language models, are prone to generating false but convincing information—a phenomenon known as "hallucinations." In nuclear contexts, this could manifest as AI systems providing confident but incorrect assessments of adversary intentions, threat levels, or appropriate responses. As researchers at War on the Rocks note, "Today's AI can confidently generate false information that can lead to flawed predictions and recommendations, ultimately skewing decision-making."

The danger is amplified by the fact that AI-generated false information often appears highly credible, potentially leading human decision-makers to base critical choices on fundamentally flawed assessments. In a nuclear crisis, an AI system might confidently report that adversary forces are preparing for attack based on misinterpreted satellite imagery or communications intercepts, potentially triggering escalatory responses based on non-existent threats.

The Black Box Problem

Current AI systems often operate as "black boxes," making decisions through processes that are opaque to human operators. This lack of transparency "undermines trust and reduces the utility of AI in high-stakes environments like nuclear decision-making, where transparency is crucial," according to analysts studying AI risks in nuclear contexts.

The opacity problem is particularly dangerous because it makes it difficult for human operators to understand why an AI system reached particular conclusions or to identify when the system might be making errors. In conventional military applications, operators might be able to verify AI assessments through other means, but nuclear crises often unfold rapidly with limited opportunities for independent verification.

Adversarial Attacks and System Manipulation

AI systems are vulnerable to sophisticated attacks designed to manipulate their decision-making processes. Adversaries could potentially "poison" the data used to train AI systems, inject false information into real-time feeds, or exploit algorithmic vulnerabilities to cause systems to reach incorrect conclusions.

In nuclear contexts, such attacks could be devastating. An adversary might manipulate AI early warning systems to generate false alarms, causing the targeted nation to escalate tensions unnecessarily. Alternatively, attackers might suppress genuine threat indicators, potentially leaving a nation vulnerable to surprise attack by degrading its AI-enhanced situational awareness.

The sophistication of such attacks is increasing rapidly. State-level adversaries possess both the technical capabilities and the motivations to develop attacks specifically designed to exploit AI vulnerabilities in nuclear systems. The challenge for defenders is that they must protect against both known attack methods and novel approaches that adversaries might develop in the future.

Strategic Escalation Risks

Beyond technical vulnerabilities, AI integration into nuclear systems could fundamentally alter strategic dynamics in ways that increase rather than decrease the risk of nuclear conflict. These strategic risks arise from the interaction between AI capabilities and the logic of nuclear deterrence.

Compressed Decision Timelines and Flash Wars

AI systems operating at machine speed could potentially accelerate crisis dynamics beyond human ability to manage them effectively. As the Arms Control Association warns, this "could create scenarios leading to 'authorized nuclear exchange' within hours" as AI systems interact with each other faster than human decision-makers can intervene.

The concept of "flash wars" driven by AI interaction has particular relevance in nuclear contexts. If multiple nations deploy AI systems to enhance their nuclear command and control, these systems might interact in ways that rapidly escalate tensions without adequate human oversight. An AI system detecting what it interprets as preparation for attack might recommend alert measures that trigger responses from adversary AI systems, creating escalation spirals that outpace human decision-making.

This risk is compounded by the fact that AI systems might optimize for speed over accuracy in crisis situations. While rapid response can be valuable in conventional conflicts, the irreversible nature of nuclear decisions means that speed should generally be subordinated to careful deliberation and verification.

Automation Bias and Human Deskilling

One of the most insidious risks of nuclear AI integration is the gradual erosion of human judgment and expertise through over-reliance on automated systems. Research on "automation bias" shows that human operators often trust automated systems even when those systems provide incorrect information.

In nuclear contexts, automation bias could prove catastrophic. As Michael Horowitz notes, "A risk in a world of automation bias is that the Petrov of the future doesn't use his judgment, or that there is no Petrov"—referring to the Soviet officer whose human judgment prevented nuclear war in 1983 when he correctly identified a false alarm from automated systems.

The concern extends beyond individual decisions to broader patterns of human deskilling. If military operators become accustomed to relying on AI for threat assessment and decision support, they might lose the independent analytical capabilities needed to question AI recommendations or function effectively when AI systems fail or are compromised.

Undermining Deterrence Logic

AI integration could potentially undermine the fundamental logic of nuclear deterrence by making "rationality-based deterrence logic increasingly untenable," as Arms Control Association researchers warn. Traditional deterrence depends on predictable, rational decision-making by human leaders who understand the consequences of nuclear use and can be influenced by threats of retaliation.

AI systems might not respond to deterrent threats in predictable ways. An AI system programmed to optimize military outcomes might recommend nuclear use in situations where human decision-makers would recognize the broader political and strategic costs. Alternatively, adversaries might believe that AI-enabled nations are more likely to use nuclear weapons quickly in crises, potentially leading them to "use or lose" calculations that increase rather than decrease nuclear risks.

Historical Precedents and Lessons

The risks of nuclear AI become more concrete when viewed through the lens of historical incidents where automated systems have malfunctioned or provided false information in nuclear contexts. These precedents illustrate how technical failures can bring the world to the brink of nuclear war.

The Stanislav Petrov Incident

The most famous example occurred on September 26, 1983, when Soviet Lieutenant Colonel Stanislav Petrov was monitoring nuclear early warning systems that detected what appeared to be incoming American missiles. The computer system "concluded with the highest confidence that the United States had launched a nuclear war," according to accounts of the incident.

However, Petrov was skeptical of the computer's assessment. He noted that the system estimated only a handful of incoming missiles, when a real American attack would likely involve an overwhelming first strike. He also distrusted the new detection system and noted that ground-based radar had not corroborated the satellite warning. Based on this human judgment, Petrov reported the alert as a false alarm, preventing Soviet retaliation that could have started nuclear war.

The incident was later determined to result from the early warning system mistaking sunlight reflecting off clouds for missile launches. As the Bulletin of Atomic Scientists notes, "If Petrov had been a machine, programmed to respond automatically when confidence was sufficiently high, that error would have started a nuclear war."

NORAD False Alarms

The United States has experienced similar incidents where automated systems generated false alarms that could have led to nuclear escalation. In November 1979, NORAD systems interpreted a missile launch drill as a real-world attack, showing hundreds of incoming Soviet missiles. The false alarm triggered alerts across multiple defense installations and nearly led to retaliatory preparations before human operators identified the error.

Additional false alarms occurred in 1980, attributed to failed computer chips and "faulty message design" in automated systems. These incidents illustrate how technical failures in automated systems can create crisis situations that require human judgment to resolve safely.

The historical pattern is clear: automated systems have repeatedly generated false positives that could have triggered nuclear responses if human operators had not exercised independent judgment to override the systems' recommendations.

Systemic and Cascading Risks

Perhaps the most concerning aspect of nuclear AI risks is their potential to create systemic vulnerabilities that cascade across multiple systems and decision-making processes. These systemic risks arise from the interconnected nature of modern nuclear command and control systems and the potential for AI failures to propagate across networks.

Classification and Information Management Challenges

AI systems operating in nuclear contexts must handle vast amounts of classified information while maintaining appropriate security protocols. However, as defense officials note, there are currently "no great answers for appropriate classification of AI products" and challenges around "classification by compilation" where AI systems might piece together classified information from unclassified sources.

These classification challenges could create operational problems where AI systems either lack access to critical information needed for accurate assessment, or alternatively, generate security risks by inappropriately handling sensitive data. In crisis situations, such information management problems could degrade decision-making quality precisely when accuracy is most critical.

Cross-Domain Interactions

Nuclear AI systems don't operate in isolation—they must integrate with conventional military systems, intelligence networks, and communications infrastructure. AI failures in one domain could potentially cascade into others, creating unexpected vulnerabilities in nuclear command and control.

For example, AI systems managing conventional forces might misinterpret nuclear-capable systems as conventional threats, potentially recommending attacks that could be interpreted as preparation for nuclear strikes. Similarly, AI-driven cyber operations might inadvertently target nuclear command systems, potentially degrading their effectiveness in crisis situations.

Alliance Coordination Challenges

AI integration also creates new challenges for alliance coordination in nuclear contexts. As NATO studies note, there are concerns that "AI systems developed by different nations may contradict each other and slow down targeting cycles rather than speeding them up." In nuclear contexts, such coordination failures could undermine extended deterrence relationships or create confusion about national decision-making authorities.

The challenge is particularly acute because allied nations are developing AI systems independently, with different technical approaches, training data, and operational concepts. Ensuring that these systems can work together effectively in nuclear crises requires unprecedented levels of technical coordination and standardization among alliance partners.

International Variations in Nuclear AI Approaches

The global landscape of nuclear AI development reveals stark differences in how various nations approach the integration of artificial intelligence into their nuclear systems. These variations reflect not only different technological capabilities but also fundamentally different strategic cultures, political systems, and threat perceptions. Understanding these national approaches is crucial for grappling with the broader implications of AI in nuclear strategy.

United States: Democratic Oversight and Controlled Integration

The United States has adopted perhaps the most cautious approach to nuclear AI integration among major powers, driven by both democratic oversight requirements and concerns about maintaining human control over ultimate strategic decisions. This approach reflects America's strategic culture of technological superiority balanced against institutional checks and accountability mechanisms.

The cornerstone of U.S. policy remains the "human in the loop" requirement established in the 2022 Nuclear Posture Review, which states that "in all cases, the United States will maintain a human 'in the loop' for all actions critical to informing and executing decisions by the President to initiate and terminate nuclear weapon employment." This policy was reinforced in the Biden Administration's 2024 Framework to Advance AI Governance and Risk Management in National Security.

However, within these constraints, U.S. military leaders are actively exploring AI applications that could enhance nuclear command and control without crossing red lines about autonomous decision-making. General Anthony Cotton, commander of U.S. Strategic Command, has been particularly vocal about AI's potential benefits, arguing that "advanced AI and robust data analytics capabilities provide decision advantage and improve our deterrence posture."

Cotton's vision focuses primarily on intelligence processing and decision support rather than autonomous weapons control. As he explained at CSIS, the goal is not "having a machine present forces to the president of the United States. It's just the opposite of that." Instead, AI would help process the enormous amounts of data required for nuclear decision-making while preserving human authority over ultimate choices.

The U.S. approach also reflects unique democratic oversight challenges. Congress has shown considerable interest in maintaining control over nuclear AI development, with the introduction of the bipartisan "Block Nuclear Launch by Autonomous AI Act" by Senator Edward Markey and Representatives Ted Lieu, Don Beyer, and Ken Buck. This legislation would codify existing Defense Department policy by ensuring that "no federal funds can be used for any launch of any nuclear weapon by an automated system without meaningful human control."

The congressional initiative reflects broader concerns about democratic accountability in AI development. As Representative Ted Lieu argued, "AI technology is developing at an extremely rapid pace" and "it is our job as Members of Congress to have responsible foresight when it comes to protecting future generations from potentially devastating consequences."

Russian Federation: The Perimeter Precedent and Autonomous Systems

Russia presents perhaps the most complex case in nuclear AI development, combining a long history of nuclear automation with contemporary AI integration efforts. The Russian approach is shaped by strategic cultures that emphasize survivability, deterrence through uncertainty, and willingness to accept risks that other nations might find unacceptable.

The most significant precedent for Russian nuclear automation is the Perimeter system, known to Western analysts as "Dead Hand." Developed during the 1980s and reportedly still operational, Perimeter represents an early form of automated nuclear command and control designed to ensure retaliation even if Soviet/Russian leadership is killed in a first strike.

According to Russian sources, Perimeter uses a network of sensors to detect nuclear explosions on Russian territory. If these sensors are triggered while the system is activated, Perimeter waits for a predetermined period for signals from senior military commanders. If no signal is received—presumably because leadership has been eliminated—the system can bypass normal command structures and delegate nuclear launch authority to relatively junior officers.

Recent reports suggest that Russia has not only maintained but upgraded Perimeter with modern artificial intelligence capabilities. A 2018 analysis described the system's "autonomous control and command system, which is based on Artificial Intelligence (AI) software" that can "accurately 'understand' that its time has come." Russian officials have indicated that the system remains "combat ready" and has been "improved" since its original deployment.

Beyond Perimeter, Russia is actively integrating AI into broader nuclear systems. The Defense Ministry's 2022 "Concept of Armed Forces Activity in Development and Use of Weapon Systems

Using AI Technologies" outlines plans for AI applications across military systems, including nuclear forces. Russian officials suggest the country is "likely to apply AI in nuclear forces command, control, management," according to U.S. intelligence assessments.

Russia's approach reflects strategic imperatives driven by perceived conventional weakness relative to NATO. As Russian analysts note, the country is using AI to "offset asymmetry in conventional power due to extensive personnel and material losses in Ukraine," with President Putin viewing AI as "critical for reducing capability gap with the West."

The Russian model demonstrates how authoritarian systems can pursue nuclear AI integration with fewer constraints than democratic states. As experts note, Russia's "authoritarian decision-making structure" enables "rapid AI integration" without the democratic oversight and public debate that characterizes Western approaches.

China: Opaque Integration and Strategic Ambiguity

China's approach to nuclear AI represents perhaps the most opaque among major powers, reflecting the country's broader strategic culture of secrecy around nuclear matters combined with rapid advancement in AI capabilities. Chinese officials have consistently refused to provide details about how AI might be integrated into nuclear systems, maintaining strategic ambiguity about both capabilities and intentions.

What is publicly known about Chinese nuclear AI comes primarily from U.S. intelligence assessments and academic analyses rather than official Chinese statements. U.S. officials indicate that "China's nuclear command and control system remains opaque, and it is not clear how the People's Liberation Army Rocket Force would employ AI in a crisis."

The opacity is intentional. Chinese officials argue that discussing nuclear AI applications with other powers "would undermine nuclear deterrence." This position reflects China's broader approach to nuclear matters, where ambiguity about capabilities and intentions is seen as enhancing rather than undermining deterrent effects.

However, China's rapid advancement in AI technologies and its stated commitment to "intelligentized warfare" suggest significant ongoing development in nuclear AI applications. Chinese military writings describe AI as central to future military competition, with particular emphasis on information dominance and "cognitive warfare" capabilities that could extend to nuclear contexts.

The most significant public indication of Chinese nuclear AI policy came in November 2024, when President Xi Jinping and U.S. President Biden jointly affirmed "the need to maintain human control over the decision to use nuclear weapons." This marked the first time China had made such a public commitment, potentially signaling recognition of nuclear AI risks.

However, the practical implementation of this commitment remains unclear. Chinese strategic culture emphasizes long-term planning and acceptance of risks for potential strategic advantages, characteristics that could support aggressive nuclear AI development despite public commitments to human control.

United Kingdom: Political Control and Alliance Coordination

The United Kingdom has adopted an approach to nuclear AI that emphasizes political rather than just military control while maintaining close coordination with alliance partners, particularly the United States. The British position reflects the country's strategic culture of balancing independence with alliance integration.

The core of UK policy is maintaining "human political control at all times" regardless of AI use in nuclear systems. This formulation differs subtly from the U.S. emphasis on "human in the loop" by specifically highlighting political rather than just human control, reflecting British constitutional traditions that emphasize civilian authority over military decisions.

British officials have indicated willingness to explore AI applications in nuclear support systems while maintaining firm boundaries around decision-making authority. The UK approach appears to parallel U.S. developments, likely reflecting close intelligence and technical cooperation between the two countries through arrangements like AUKUS.

The British position also reflects practical constraints. As a smaller nuclear power with limited resources for independent development, the UK likely relies heavily on cooperation with the United States for advanced nuclear AI capabilities. This dependency creates both opportunities for alliance standardization and risks about maintaining independent decision-making capabilities.

Emerging Nuclear Powers: Varied Approaches

Smaller nuclear powers are approaching nuclear AI with varying levels of sophistication and different strategic priorities. These approaches often reflect resource constraints, threat perceptions, and relationships with major powers.

India has demonstrated growing interest in military AI applications but has maintained relative silence about nuclear AI specifically. India's strategic culture emphasizes strategic autonomy and gradual capability development, suggesting a cautious approach to nuclear AI integration that prioritizes maintaining independent decision-making capabilities.

Pakistan faces particular challenges in nuclear AI development due to resource constraints and security concerns. The country's nuclear doctrine emphasizes delegation of authority to military commanders in crisis situations, potentially creating tensions with requirements for human control over AI-enabled systems.

Israel has demonstrated advanced capabilities in military AI but maintains its traditional policy of nuclear opacity, making it difficult to assess specific nuclear AI developments. Israel's strategic culture emphasizes technological superiority and rapid adaptation, characteristics that could support aggressive nuclear AI development.

North Korea represents perhaps the most unpredictable case, combining limited technical capabilities with willingness to accept extreme risks. North Korean nuclear AI development is likely limited by sanctions and resource constraints, but the regime's risk tolerance could lead to deployment of unreliable systems that other nations would reject.

Implications for Strategic Stability

The varied national approaches to nuclear AI create several challenges for strategic stability. Different technical standards, operational concepts, and command structures could complicate crisis management and arms control verification. Nations developing AI systems independently may create incompatible approaches that increase rather than decrease nuclear risks.

The lack of transparency, particularly from China and Russia, makes it difficult for other nations to assess nuclear AI capabilities and adjust their own policies accordingly. This uncertainty could drive worst-case planning and accelerate AI development in ways that undermine stability.

Most concerning is the potential for a "lowest common denominator" effect, where the most aggressive approach to nuclear AI integration drives other nations to match or exceed those capabilities. If one nation deploys AI systems that provide significant decision-making speed advantages, others may feel compelled to follow suit despite safety concerns.

Emerging Concepts and Strategic Implications

The integration of artificial intelligence into nuclear systems is generating entirely new strategic concepts that challenge traditional assumptions about deterrence, escalation, and crisis management. These emerging ideas represent fundamental shifts in how military planners and strategists think about nuclear weapons in the digital age, with implications that extend far beyond technical questions of system design.

Algorithmic Escalation Dynamics

Perhaps the most significant emerging concept is "algorithmic escalation"—the possibility that AI systems could accelerate crisis dynamics beyond human ability to manage them effectively. This represents a fundamental departure from traditional escalation theory, which assumes human decision-makers will have adequate time to assess situations, consult advisors, and choose appropriate responses.

Algorithmic escalation could manifest in several ways. AI systems designed to optimize military outcomes might recommend rapid escalation based on narrow technical criteria without considering broader political and strategic factors that human decision-makers would take into account. Alternatively, AI systems from different nations might interact with each other in ways that create escalation spirals, with each system responding to the other's moves faster than human operators can intervene.

The speed differential is crucial. While human decision-making in nuclear crises typically unfolds over hours or days, AI systems can potentially assess situations and recommend responses in seconds or minutes. This compression of decision timelines could fundamentally alter crisis dynamics, reducing opportunities for diplomacy, consultation, and careful deliberation that have historically helped prevent nuclear escalation.

Recent studies suggest that this risk is not merely theoretical. As noted in Arms Control Association research, AI systems could create "crisis scenarios where tensions are high and communications are strained" leading to "authorized nuclear exchange" within hours as systems interact faster than human decision-makers can control them.

Decision Compression and Strategic Patience

The concept of "decision compression" represents another fundamental challenge to traditional nuclear strategy. Nuclear deterrence has historically relied on what strategists call "strategic patience"—the ability of decision-makers to resist pressure for immediate action and instead pursue deliberate, measured responses that avoid unnecessary escalation.

AI integration threatens this strategic patience by creating pressure for rapid responses that match the speed of automated systems. If adversaries deploy AI systems that can assess threats and recommend responses in minutes, traditional approaches that rely on hours or days of deliberation may appear too slow to be effective.

However, this emphasis on speed may be fundamentally misguided in nuclear contexts. Unlike conventional warfare, where rapid response can provide tactical advantages, nuclear weapons are primarily political instruments whose effectiveness depends more on the credibility of threatened retaliation than on the speed of that retaliation.

The challenge is that AI systems might not distinguish between contexts where speed is valuable and those where deliberation is more important. An AI system trained on conventional warfare data might recommend rapid nuclear responses in situations where human decision-makers would recognize the need for careful consideration and diplomatic communication.

Autonomous Nuclear Platforms and Strategic Implications

Russia's development of systems like the Poseidon nuclear-powered, nuclear-armed underwater drone represents another emerging concept: autonomous nuclear platforms that can operate independently for extended periods without direct human control. These systems challenge traditional assumptions about human oversight of nuclear weapons by creating platforms that may be physically unable to receive human commands once deployed.

The strategic implications are profound. Autonomous nuclear platforms could potentially continue operating even after their launching nation's command and control systems are destroyed, creating scenarios where nuclear weapons remain active without any human authority to control or recall them. This could fundamentally alter post-attack scenarios and complicate efforts to terminate nuclear conflicts.

Moreover, autonomous platforms raise questions about the meaning of "human control" in nuclear systems. While a human decision-maker might authorize the initial deployment of an autonomous nuclear platform, subsequent decisions about targeting and weapons use might be delegated to AI systems operating beyond human oversight.

The proliferation of such systems could create new categories of nuclear risk. Multiple autonomous platforms operating simultaneously might interact in unpredictable ways, potentially creating escalation scenarios that no human decision-maker can effectively control or terminate.

Information Warfare and Nuclear Signaling

AI is also transforming the information environment in which nuclear decisions are made, creating new possibilities for both deterrence and miscalculation. AI-powered information operations could potentially manipulate the data that feeds into nuclear decision-making systems, creating false impressions of adversary intentions or capabilities.

This has particular relevance for nuclear signaling—the complex communication process through which nations communicate resolve, reassurance, and red lines to potential adversaries. Traditional signaling relies on predictable patterns of military deployment, diplomatic communication, and public statements that adversaries can interpret accurately.

AI systems might disrupt these signaling processes in several ways. They could generate false signals that mislead adversaries about intentions and capabilities, or they might misinterpret genuine signals in ways that lead to inappropriate responses. The challenge is that effective nuclear deterrence depends on clear communication between adversaries, and AI systems might introduce noise or distortion into these critical communication channels.

Cross-Domain Entanglement

The concept of "cross-domain entanglement" describes how AI systems operating in different domains—cyber, space, conventional forces, and nuclear systems—might become interconnected in ways that create unexpected escalation pathways. This represents a departure from traditional approaches that treat nuclear and conventional systems as largely separate domains.

AI integration increases entanglement risks because AI systems often rely on shared data sources, communications networks, and decision-making processes. An attack on conventional AI systems might inadvertently affect nuclear systems, or AI-driven conventional operations might be misinterpreted as preparation for nuclear attacks.

For example, AI systems managing conventional cyber operations might target adversary communications networks that also support nuclear command and control. While the attacking nation might view this as routine conventional warfare, the targeted nation might interpret disruption of its nuclear communications as preparation for nuclear attack.

Deterrence Theory Adaptation

These emerging concepts are forcing fundamental reconsideration of deterrence theory for the AI age. Traditional deterrence theory assumes rational actors who can accurately assess threats, communicate intentions clearly, and maintain control over escalation processes. AI integration challenges all of these assumptions.

The concept of "rational deterrence" becomes problematic when AI systems might optimize for objectives that human decision-makers would consider irrational in broader strategic contexts. An AI system programmed to maximize military effectiveness might recommend nuclear use in situations where human decision-makers would recognize the devastating political and strategic costs.

Similarly, the predictability that underpins deterrence relationships could be undermined by AI systems whose decision-making processes are opaque or whose training data leads them to conclusions that human adversaries cannot anticipate. Effective deterrence requires that potential adversaries can predict how their actions will be interpreted and what responses they might provoke.

The Singularity Problem in Nuclear Strategy

Some analysts have begun discussing the possibility of a "nuclear singularity"—a point at which AI systems become so sophisticated and operate so quickly that human beings can no longer meaningfully control nuclear decisions. This concept, adapted from broader discussions about artificial general intelligence, suggests that AI development might eventually reach a point where human oversight becomes impossible rather than merely difficult.

The nuclear singularity concept raises fundamental questions about the future of human agency in nuclear strategy. If AI systems become capable of making strategic decisions that humans cannot understand or predict, traditional concepts of command authority, democratic control, and political accountability might become meaningless.

While current AI systems are far from reaching such capabilities, the rapid pace of AI development suggests that these questions may become practical rather than theoretical within decades. The challenge for current policymakers is developing governance frameworks that can adapt to dramatically different AI capabilities while preserving essential human control over nuclear decisions.

Strategic Stability in the AI Age

All of these emerging concepts contribute to growing uncertainty about strategic stability in an AI-enabled world. Traditional arms control approaches, which focus on limiting numbers and types of weapons, may be inadequate for addressing the risks created by AI integration into nuclear systems.

The challenge is that AI capabilities are often software-based and therefore difficult to verify or limit through traditional arms control mechanisms. Nations might agree to maintain human control over nuclear decisions while secretly developing AI systems that effectively automate those decisions. Similarly, AI systems that appear to be defensive early warning systems might actually have offensive capabilities that are difficult to detect.

Moreover, the speed at which AI systems can operate might outpace traditional crisis management mechanisms. Arms control agreements and crisis communication protocols developed for human decision-makers might be too slow to prevent AI-driven escalation scenarios.

The emerging consensus among experts is that managing nuclear risks in the AI age will require new approaches that go beyond traditional arms control to address the underlying technologies and decision-making processes that create these risks. This might include technical standards for AI systems in nuclear applications, transparency requirements for AI decision-making processes, and new forms of verification that can address software-based capabilities.

Policy Responses and Governance Frameworks

The recognition of nuclear AI risks has prompted a range of policy responses from national governments, international organizations, and civil society groups. These initiatives represent attempts to develop governance frameworks that can harness the potential benefits of AI in nuclear systems while preventing catastrophic risks. However, the rapid pace of AI development and the sensitive nature of nuclear systems create significant challenges for effective governance.

Legislative Responses and Congressional Action

The most concrete policy response to nuclear AI risks has come from the U.S. Congress, which has introduced bipartisan legislation specifically designed to prevent autonomous nuclear weapons launches. The "Block Nuclear Launch by Autonomous AI Act," introduced by Senator Edward Markey and Representatives Ted Lieu, Don Beyer, and Ken Buck, represents the first comprehensive legislative attempt to address nuclear AI risks.

The legislation would codify existing Defense Department policy by ensuring that "no federal funds can be used for any launch of any nuclear weapon by an automated system without meaningful human control." The bill goes beyond current policy by creating legal requirements rather than just administrative guidelines, making it more difficult for future administrations to modify policies without congressional approval.

The legislative language is carefully crafted to address specific risks while allowing continued development of AI support systems. The bill prohibits autonomous launch decisions but permits AI applications in intelligence processing, threat assessment, and decision support—maintaining the distinction between AI that assists human decision-making and AI that replaces it.

Supporting the legislation, Senator Markey argued that "as we live in an increasingly digital age, we need to ensure that humans hold the power alone to command, control, and launch nuclear weapons—not robots." Representative Lieu emphasized the forward-looking nature of the legislation: "It is our job as Members of Congress to have responsible foresight when it comes to protecting future generations from potentially devastating consequences."

The bipartisan nature of the legislation is significant, with co-sponsors including Senators Bernie Sanders, Elizabeth Warren, and Jeff Merkley, along with Republican Representative Ken Buck. This broad support reflects growing recognition that nuclear AI risks transcend traditional partisan divisions.

However, the legislation also faces criticism from some defense analysts who argue that rigid restrictions on AI applications could handicap the United States relative to adversaries who might be less constrained. Critics worry that the bill's language might be interpreted too broadly, potentially limiting beneficial AI applications in nuclear support systems.

Executive Branch Policy Development

The Biden Administration developed several policy frameworks for addressing AI in national security contexts, including nuclear systems. The October 2024 "Framework to Advance AI Governance and Risk Management in National Security" specifically reaffirmed the human-in-the-loop requirement for nuclear decisions while providing guidance for AI development in other military applications.

The framework establishes several key principles for nuclear AI development:

- Maintaining "meaningful human control" over all nuclear employment decisions
- Requiring rigorous testing and validation of AI systems before deployment
- Ensuring transparency and explainability in AI decision-making processes
- Establishing clear accountability mechanisms for AI system failures
- Mandating regular reviews of AI system performance and safety

However, implementation of these principles faces significant challenges. As defense officials acknowledge, there are currently "no great answers for appropriate classification of AI products" in nuclear contexts, making it difficult to establish clear guidelines for development and deployment.

The classification challenge is particularly acute because AI systems often require access to vast amounts of data to function effectively, but nuclear systems involve some of the most sensitive information in government. Balancing AI capability requirements with security needs requires new approaches to information management and system design.

Department of Defense Implementation

The Pentagon has begun developing specific guidance for AI applications in nuclear systems, building on broader AI adoption policies while recognizing the unique requirements of nuclear command and control. The Defense Department's approach emphasizes gradual, carefully controlled integration that maintains human decision-making authority.

General Anthony Cotton, commander of U.S. Strategic Command, has outlined a vision for nuclear AI that focuses on enhancing human decision-making rather than replacing it. As he explained to Congress, "AI will enhance our decision-making capabilities. But we must never allow artificial intelligence to make those decisions for us."

The Strategic Command approach involves several specific applications:

- Using AI to process intelligence data more rapidly and accurately
- Employing machine learning to improve early warning system performance
- Developing AI-assisted planning tools that help commanders evaluate response options
- Implementing AI-powered communications and system management capabilities

However, all these applications must operate within strict boundaries that preserve human control over ultimate decisions. As Cotton emphasized, the goal is "not having a machine present forces to the president of the United States. It's just the opposite of that."

The Pentagon is also developing new testing and validation procedures specifically for AI systems in nuclear contexts. These procedures must address the unique challenge that nuclear AI systems cannot be tested in realistic scenarios, requiring extensive simulation and modeling to ensure reliability.

International Cooperation and Diplomatic Initiatives

Recognition that nuclear AI risks are inherently global has prompted various international cooperation efforts, though progress has been limited by the sensitive nature of nuclear systems and competitive dynamics among major powers.

The most significant breakthrough came in November 2024 when Presidents Biden and Xi Jinping jointly affirmed "the need to maintain human control over the decision to use nuclear weapons." This marked the first time China had made such a public commitment, representing a potential foundation for broader international cooperation on nuclear AI governance.

The U.S.-China statement built on earlier initiatives, including a 2022 joint declaration by France, the United Kingdom, and the United States during the Nuclear Non-Proliferation Treaty review process. These nations committed to maintaining human control over nuclear decisions while exploring AI applications in supporting roles.

However, translating these high-level commitments into practical cooperation faces significant obstacles. Nations are reluctant to share information about nuclear AI development that could reveal capabilities or vulnerabilities. Moreover, verification of compliance with human control commitments is extremely difficult given the software-based nature of many AI capabilities.

The lack of Russian participation in most international AI governance initiatives represents a particular challenge. Russia's development of systems like Perimeter suggests willingness to accept higher levels of automation in nuclear systems than other major powers consider appropriate.

United Nations and Multilateral Framework Development

The United Nations has attempted to address nuclear AI risks through broader initiatives on lethal autonomous weapons systems, though progress has been limited by disagreements among member states about appropriate restrictions.

The UN General Assembly's December 2024 resolution on lethal autonomous weapons systems, which passed with 166 votes in favor, acknowledged the "serious challenges and concerns" raised by "artificial intelligence and autonomy in weapons systems." However, the resolution focused primarily on conventional weapons rather than nuclear systems specifically.

UN Secretary-General António Guterres has been particularly vocal about nuclear AI risks, stating that nuclear weapons should "always include humans within decision-making chains." In his 2023 "New Agenda for Peace," Guterres called for member states to conclude by 2026 "a legally binding instrument to prohibit lethal autonomous weapon systems" that could extend to nuclear applications.

The Convention on Certain Conventional Weapons (CCW) has provided a forum for discussing autonomous weapons issues, including nuclear applications. However, the CCW's consensus requirement has limited progress, with major military powers blocking proposals for binding restrictions on autonomous weapons systems.

Technical Standards and Verification Challenges

Developing effective governance for nuclear AI requires addressing fundamental technical challenges around standards, testing, and verification. Unlike traditional arms control, which focuses on physical systems that can be counted and monitored, AI governance must address software capabilities that may be difficult to detect or verify.

Several organizations are working to develop technical standards for AI in nuclear applications:

IEEE Standards Development: The Institute of Electrical and Electronics Engineers is developing standards for AI system reliability, transparency, and human oversight that could apply to nuclear systems.

NATO Standardization: The alliance is working on common technical requirements for AI systems that must operate together in crisis situations, including potential nuclear scenarios.

Industry Self-Regulation: Major AI companies are developing internal standards for AI applications in sensitive domains, though these voluntary approaches may be insufficient for nuclear applications.

However, verification of compliance with technical standards remains extremely challenging. AI systems can be modified rapidly through software updates, and their decision-making processes may be difficult to audit comprehensively. Traditional arms control verification mechanisms are poorly suited to these characteristics.

Civil Society and Academic Contributions

Non-governmental organizations and academic institutions have played important roles in developing policy frameworks for nuclear AI governance. These contributions often provide technical expertise and policy analysis that governments may lack, particularly given the rapid pace of AI development.

The Arms Control Association has published detailed analyses of nuclear AI risks and proposed specific policy responses, including technical safeguards and verification mechanisms. The organization's work has informed legislative proposals and executive branch policy development.

Academic institutions like the Center for Strategic and International Studies and the Carnegie Endowment for International Peace have provided forums for dialogue between government officials, industry representatives, and civil society groups. These discussions help build consensus around appropriate governance approaches while identifying areas of continued disagreement.

The Bulletin of Atomic Scientists has been particularly active in highlighting nuclear AI risks through its Doomsday Clock and related publications. The organization's emphasis on long-term existential risks provides important perspective on nuclear AI issues that might otherwise focus too narrowly on near-term technical challenges.

Implementation Challenges and Future Directions

Despite growing recognition of nuclear AI risks and various policy initiatives, significant challenges remain in developing effective governance frameworks. These challenges reflect both technical limitations and political obstacles that constrain policy development.

Technical Complexity: Nuclear AI systems involve extremely complex interactions between hardware, software, and human operators. Developing governance frameworks that address this complexity without stifling beneficial innovation requires expertise that may not exist in traditional policy communities.

Classification and Transparency: The classified nature of nuclear systems makes it difficult to develop governance frameworks that require transparency about AI capabilities and decision-making processes. Balancing security requirements with governance needs remains an unsolved challenge.

International Coordination: Effective nuclear AI governance requires cooperation among nations that view each other as adversaries in other domains. Building trust and verification mechanisms in this environment requires diplomatic approaches that may not yet exist.

Rapid Technological Change: The pace of AI development means that governance frameworks may become obsolete before they can be fully implemented. Creating adaptive governance mechanisms that can evolve with technology requires new approaches to policy development.

The path forward likely requires continued development of multiple complementary approaches: domestic legislation and regulation, international cooperation agreements, technical standards development, and civil society monitoring and advocacy. Success will depend on sustained attention to these issues even as other priorities compete for policymaker attention.

Key Insight: The Ultimate Test of Human-AI Integration

The debate over artificial intelligence in nuclear strategy represents far more than a technical discussion about system capabilities or a policy argument about appropriate safeguards. It embodies the ultimate test of how strategic cultures can adapt to the AI revolution while preserving what is most essential about human agency in matters of war and peace. Nuclear strategy in the age of algorithms forces us to confront fundamental questions about the relationship between human wisdom and machine intelligence in contexts where the stakes could not be higher.

This challenge is unique in the history of military technology because nuclear weapons occupy a special place in strategic thinking—they are simultaneously military instruments and political symbols, tactical capabilities and existential threats. Unlike other military technologies that can be tested, refined, and gradually improved through battlefield experience, nuclear weapons must work correctly in their most important application—deterring their own use. This creates extraordinary demands for reliability, predictability, and human control that may be fundamentally incompatible with the current state of AI technology.

The Irreversibility Problem

The most fundamental challenge nuclear strategy poses for AI integration is what we might call the "irreversibility problem." Most military decisions, even those with significant consequences, can be modified, corrected, or reversed as circumstances change. Nuclear decisions, once implemented, cannot be undone. A nuclear weapon, once launched, cannot be recalled. A nuclear exchange, once begun, cannot be rewound. This irreversibility creates unique requirements for decision-making processes that must be correct not just most of the time, but essentially all of the time.

Current AI systems, no matter how sophisticated, operate through probabilistic processes that involve uncertainty and error rates. Machine learning algorithms improve their performance over time by learning from mistakes. But nuclear strategy cannot accommodate learning from mistakes about nuclear use—the consequences of algorithmic errors in nuclear contexts could be civilization-ending. This creates a fundamental mismatch between how AI systems naturally function and what nuclear strategy requires.

The irreversibility problem also extends to the broader strategic environment. If nations begin integrating AI into nuclear systems in ways that prove destabilizing or dangerous, reversing those integrations may be extremely difficult. AI capabilities, once developed and deployed, create pressures for other nations to match or exceed them. The result could be irreversible changes to the strategic environment that make nuclear conflict more rather than less likely.

The Context and Judgment Challenge

Nuclear decision-making inherently involves forms of context and judgment that may be beyond current AI capabilities. Nuclear weapons are not just military tools but political instruments whose use carries moral, legal, and strategic implications that extend far beyond immediate military objectives. Deciding whether, when, and how to use nuclear weapons requires understanding complex political dynamics, alliance relationships, moral considerations, and long-term strategic consequences that may be difficult to quantify in ways that AI systems can process.

The Stanislav Petrov incident illustrates this challenge perfectly. When Soviet early warning systems detected what appeared to be incoming American missiles in 1983, Petrov's decision to report the alert as a false alarm was based not just on technical analysis but on strategic intuition—his judgment that a real American attack would likely involve many more missiles, and his skepticism about the reliability of the new detection system. This kind of contextual reasoning, combining technical knowledge with strategic understanding and appropriate skepticism about automated systems, represents exactly the kind of human judgment that nuclear decisions require.

An AI system trained on the same data might have reached a different conclusion. If programmed to respond to high-confidence threat detections, it might have recommended immediate retaliation based on the system's assessment. If trained to optimize military outcomes, it might not have weighed the broader strategic considerations that informed Petrov's human judgment. The result could have been nuclear war based on a false alarm—precisely the kind of catastrophic error that nuclear strategy must prevent.

The Democratic Accountability Dilemma

For democratic nations, nuclear AI integration raises profound questions about accountability and civilian control that go to the heart of democratic governance. Democratic theory requires that the most consequential decisions—particularly those involving war and peace—be subject to civilian oversight and public accountability. Nuclear weapons represent the ultimate expression of state power, making their control a fundamental question of democratic governance.

AI integration threatens this accountability in several ways. First, the complexity and opacity of AI systems make it difficult for civilian officials to understand how these systems reach their conclusions or to exercise meaningful oversight over their operations. Second, the speed at which AI systems operate may compress decision timelines in ways that prevent adequate consultation with civilian authorities or deliberation about appropriate responses.

Perhaps most concerning, AI systems might gradually shift decision-making authority away from elected officials toward technical experts and military operators who understand the systems but may lack the political authority to make strategic decisions. This could represent a subtle but significant erosion of democratic control over nuclear weapons, occurring not through any single dramatic decision but through incremental technical choices that cumulatively alter the balance of authority.

The Congressional response to these concerns, embodied in the "Block Nuclear Launch by Autonomous AI Act," reflects recognition that preserving democratic control over nuclear decisions requires active legislative intervention rather than simply trusting executive branch officials to maintain appropriate boundaries. The bipartisan nature of this legislation suggests broad recognition that nuclear AI raises questions that transcend normal partisan divisions.

The Alliance and Strategic Culture Challenge

Nuclear AI integration also poses unique challenges for alliance relationships and strategic culture transmission. Nuclear deterrence often depends on extended deterrence relationships where one nation's nuclear weapons protect allies who do not possess their own nuclear capabilities. These relationships require high levels of trust and coordination that could be undermined by AI integration in several ways.

First, if nuclear decision-making becomes increasingly automated, allies may lose confidence that their protector will make appropriate decisions in crisis situations. The credibility of extended deterrence depends partly on allies' belief that the protecting nation will make rational, measured decisions that take alliance interests into account. AI systems that optimize for narrow military or technical criteria might not provide this assurance.

Second, AI integration might create technical barriers to alliance coordination. If nations develop AI systems with different capabilities, training data, or operational concepts, these systems might not be able to coordinate effectively in crisis situations. Worse, they might actively conflict with each other, creating confusion and miscommunication precisely when clarity and coordination are most important.

Finally, AI integration raises questions about strategic culture transmission across generations. Strategic culture—the deeply held beliefs about how to approach security challenges—is typically transmitted through professional military education, operational experience, and mentorship relationships. If AI systems increasingly handle the analytical and planning functions that have traditionally taught military officers how to think strategically, future generations of military leaders might lose essential capabilities for independent strategic judgment.

The Ultimate Test of Human-AI Partnership

Despite these challenges, the nuclear AI debate also represents an opportunity to develop models of human-AI partnership that could have broader applications in other high-stakes domains. If the national security community can successfully develop approaches to nuclear AI that preserve human agency while leveraging machine capabilities, these models could inform AI governance in other areas where the consequences of errors are severe.

The key is recognizing that successful human-AI partnership in nuclear contexts cannot simply involve humans supervising AI systems or AI systems supporting human decisions. Instead, it requires fundamental rethinking of how humans and machines can work together in ways that combine the unique strengths of each while compensating for their respective limitations.

Human beings bring to nuclear decision-making capabilities that current AI systems cannot match: moral reasoning, political judgment, creative problem-solving, and the ability to understand context and meaning in ways that go beyond pattern recognition. AI systems bring speed, computational power, and the ability to process vast amounts of data without being overwhelmed by stress or emotional pressure.

Successful integration would require designing systems that leverage AI's computational capabilities while preserving space for human judgment about the most important questions. This might involve AI systems that can rapidly process intelligence data and identify potential threats, but that present their findings to human decision-makers in ways that preserve space for contextual judgment and appropriate skepticism about machine recommendations.

The Path Forward: Conscious Design of Strategic Culture

The nuclear AI challenge ultimately requires conscious design of strategic cultures that can function effectively in human-AI partnership models. This goes beyond simply adding AI capabilities to existing strategic cultures—it requires fundamental rethinking of how strategic cultures can preserve their essential characteristics while adapting to new technological realities.

Strategic cultures have always evolved in response to technological change, but the pace and scope of AI development may require more conscious and deliberate cultural evolution than previous technologies have demanded. Rather than allowing strategic cultures to adapt organically to AI capabilities, leaders may need to actively design cultural frameworks that can guide human-AI interaction in nuclear contexts.

This conscious design process must address several key questions: How can strategic cultures maintain appropriate skepticism about machine recommendations while still benefiting from AI capabilities? How can they preserve space for moral reasoning and political judgment in an environment increasingly dominated by technical analysis? How can they maintain democratic accountability and alliance coordination while adapting to the speed requirements of AI-enabled warfare?

The answers to these questions will shape not just nuclear strategy but the broader relationship between human agency and artificial intelligence in strategic affairs. Nuclear strategy, because of its unique characteristics and ultimate stakes, provides the most demanding test case for these relationships. Success in meeting this test could provide models for human-AI partnership that enhance rather than undermine human agency in other strategic domains.

Failure, however, could result in the gradual erosion of human control over the most consequential decisions any society can make. The nuclear AI debate thus represents far more than a technical or policy challenge—it embodies the fundamental question of whether human beings can remain in control of their own destiny in an age of increasingly powerful artificial intelligence. The stakes, quite literally, could not be higher.

Source References

1. CSIS. (2025, January 24). "PONI Live Debate: AI Integration in NC3." *Center for Strategic and International Studies*. https://www.csis.org/analysis/poni-live-debate-ai-integration-nc3
2. Markey, E. J., Lieu, T. W., Beyer, D. S., & Buck, K. (2023, April 26). "Markey, Lieu, Beyer, and Buck Introduce Bipartisan Legislation to Prevent AI From Launching a Nuclear Weapon." *U.S. Senator Ed Markey Press Release*. https://www.markey.senate.gov/news/press-releases/markey-lieu-beyer-and-buck-introduce-bipartisan-legislation-to-prevent-ai-from-launching-a-nuclear-weapon
3. Freedberg Jr., S. J. (2025, January 24). "ChatNC3: Can the US trust AI in nuclear command, control and communications?" *Breaking Defense*. https://breakingdefense.com/2025/01/chatnc3-can-the-us-trust-ai-in-nuclear-command-control-and-communications/
4. Williams, H. (2024, December 10). "Updating Nuclear Command, Control, and Communication." *Center for Strategic and International Studies*. https://www.csis.org/analysis/updating-nuclear-command-control-and-communication
5. Scharre, P., & Others. (2024, January 17). "Artificial Intelligence and Nuclear Stability." *War on the Rocks*. https://warontherocks.com/2024/01/artificial-intelligence-and-nuclear-stability/
6. Hadley, G. (2024, October 28). "STRATCOM Boss: AI 'Will Enhance' Nuclear C2." *Air & Space Forces Magazine*. https://www.airandspaceforces.com/stratcom-boss-ai-nuclear-command-control/
7. Gertz, B. (2024, August 13). "U.S. will not let automated AI system decide on nuclear strikes, top general says." *Washington Times*. https://www.washingtontimes.com/news/2024/aug/13/us-will-not-let-automated-ai-system-decide-on-nucl/
8. Bulletin of Atomic Scientists. (2022, February 15). "Giving an AI control of nuclear weapons: What could possibly go wrong?" https://thebulletin.org/2022/02/giving-an-ai-control-of-nuclear-weapons-what-could-possibly-go-wrong/
9. Arms Control Association. (2022, September). "On Integrating Artificial Intelligence With Nuclear Control." https://www.armscontrol.org/act/2022-09/features/integrating-artificial-intelligence-nuclear-control
10. O'Hanlon, M. E. (2025, February 28). "How unchecked AI could trigger a nuclear war." *Brookings Institution*. https://www.brookings.edu/articles/how-unchecked-ai-could-trigger-a-nuclear-war/
11. Wikipedia. (2024). "1983 Soviet nuclear false alarm incident." https://en.wikipedia.org/wiki/1983_Soviet_nuclear_false_alarm_incident
12. UPI. (2019, November 8). "False alarm: 1979 NORAD incident was one of several Cold War nuclear scares." https://www.upi.com/Top_News/US/2019/11/08/False-alarm-1979-

NORAD-scare-was-one-of-several-nuclear-close-calls/7491573181627/

13. RAND Corporation. (2018, April 24). "How Artificial Intelligence Could Increase the Risk of Nuclear War." https://www.rand.org/pubs/articles/2018/how-artificial-intelligence-could-increase-the-risk.html
14. UN Office for Disarmament Affairs. (2024). "Lethal Autonomous Weapon Systems (LAWS) – UNODA." https://disarmament.unoda.org/the-convention-on-certain-conventional-weapons/background-on-laws-in-the-ccw/
15. Human Rights Watch. (2024, January 3). "Killer Robots: UN Vote Should Spur Action on Treaty." https://www.hrw.org/news/2024/01/03/killer-robots-un-vote-should-spur-action-treaty
16. Arms Control Association. (2024, November 12). "Beyond a Human 'In the Loop': Strategic Stability and Artificial Intelligence." https://www.armscontrol.org/issue-briefs/2024-011/beyond-the-loop
17. War on the Rocks. (2024, December 5). "Beyond Human-in-the-Loop: Managing AI Risks in Nuclear Command-and-Control." https://warontherocks.com/2024/12/beyond-human-in-the-loop-managing-ai-risks-in-nuclear-command-and-control/
18. The War Zone. (2025, March 7). "How The Military Wants AI To Help Control America's Nuclear Arsenal." https://www.twz.com/nuclear/this-is-how-the-military-wants-ai-to-help-control-americas-nuclear-arsenal
19. CNAS. (2025, February 14). "The AI-Nuclear Nexus: New CNAS Report on Managing Artificial Intelligence and Nuclear Weapons amid U.S. Rivalry with China and Russia." https://www.cnas.org/press/press-release/the-ai-nuclear-nexus-new-cnas-report-on-managing-artificial-intelligence-and-nuclear-weapons-amid-u-s-rivalry-with-china-and-russia
20. Union of Concerned Scientists. (2024, March 4). "Artificial Intelligence and the Evolving Landscape of Nuclear Strategy." https://blog.ucs.org/science-blogger/artificial-intelligence-and-the-evolving-landscape-of-nuclear-strategy

Chapter 10: Cyber, Space, and Information Domains: The New High Ground

AI Integration Across Non-Kinetic Battlespaces

The integration room at Taiwan's Central Epidemic Command Center hummed with activity in early 2024, but not from any public health crisis. Multiple screens displayed real-time feeds of social media platforms, satellite imagery, and network traffic analysis as teams of analysts worked to counter what officials described as the most sophisticated information warfare campaign in the island's history. Chinese cognitive warfare operations were targeting Taiwan's electoral process using artificial intelligence to "manipulate enemy perceptions and control information flows" simultaneously across cyber, space, and information spheres.

The operation demonstrated how modern strategic competition had evolved beyond traditional kinetic threats. Chinese forces deployed "sophisticated machine learning models to create personalized propaganda aimed at Taiwan," according to intelligence assessments, while simultaneously conducting cyberattacks and using space-based assets for surveillance and communications interference. The campaign represented a convergence that military strategists are only beginning to understand: AI-enabled operations that transcend traditional domain boundaries and create entirely new forms of strategic leverage.

During Taiwan's January 2024 presidential election, artificial intelligence became the invisible battlefield where democratic legitimacy itself was under assault. AI-generated deepfake videos promoted hoaxes such as a "secret history" of outgoing President Tsai Ing-wen, while coordinated bot networks amplified divisive content across multiple platforms. Microsoft's threat intelligence team later confirmed this as "the first confirmed use of AI-generated material by a nation-state to influence a foreign election," marking a watershed moment in the weaponization of artificial intelligence for strategic objectives.

The attacks were not limited to the information domain. Chinese cyber units simultaneously targeted Taiwan's critical infrastructure through what analysts describe as "cognitive electronic warfare," using AI to enhance traditional cyber operations while space-based assets provided coordination and command capabilities. The multi-domain approach revealed how AI was erasing the boundaries between cyber, space, and information operations, creating what Chinese military strategists call "intelligentized warfare"—a concept where victory comes through dominating the cognitive battlespace rather than physical terrain.

This convergence represents more than just technological advancement; it signals a fundamental transformation in how nations compete for strategic advantage. Where previous conflicts were defined by control of geographic domains—land, sea, air—the new competition centers on information superiority achieved through AI-enabled operations that span multiple domains simultaneously. The implications extend far beyond Taiwan, as similar techniques are being refined and deployed across global flashpoints from Eastern Europe to the South China Sea.

Domain Convergence Through AI

The traditional military conception of distinct operational domains—land, sea, air, space, and cyber—is rapidly becoming obsolete. Artificial intelligence is acting as a unifying force that enables simultaneous operations across these domains, creating what defense analysts describe as "multi-domain operations" but what is actually something more fundamental: the emergence of a single, integrated battlespace where the boundaries between domains exist only in outdated organizational charts.

Information Operations Revolution

The most visible manifestation of this convergence is in information warfare, where AI has transformed both the scale and sophistication of operations. China's approach to "cognitive warfare" exemplifies this transformation. As outlined in People's Liberation Army doctrine, cognitive warfare represents "the systematic use of artificial intelligence, big data analytics, and psychological operations to manipulate enemy perceptions, degrade decision-making capabilities, and control information flows before and during conflict."

The Chinese deployment of AI-enhanced psychological operations goes far beyond traditional propaganda. Intelligence assessments reveal that Chinese forces have developed "sophisticated machine learning models to create personalized propaganda aimed at swaying opinions in Taiwan and other key regions." These systems can generate thousands of variations of the same message, each tailored to specific demographic groups, psychological profiles, and even individual users based on their social media activity and behavioral patterns.

The scale of these operations has expanded exponentially with AI automation. During the 2024 Romanian presidential election, intelligence agencies documented evidence of algorithmic manipulation involving 25,000 TikTok accounts, some dormant since 2016, that suddenly became active to support pro-Russian candidate Călin Georgescu. The engagement data displayed "clear patterns of artificial manipulation, involving bot-driven activity and coordinated amplification efforts," according to declassified intelligence reports. Romania's Constitutional Court ultimately annulled the election results—an unprecedented decision that underscored how AI-enabled information warfare now poses existential threats to democratic processes.

The evolution from traditional disinformation to AI-generated content represents a qualitative leap in capability. Where previous information operations required human content creators and were limited by linguistic and cultural barriers, AI systems can now generate content in any language, adapted to local cultural contexts, and distributed at machine speed across global networks. The result is what researchers describe as "memetic warfare" where AI-generated memes, videos, and narratives can go viral and shape public opinion faster than traditional fact-checking mechanisms can respond.

Cyber-AI Integration

The cybersecurity landscape has been fundamentally altered by AI integration, creating new capabilities for both offensive and defensive operations. The U.S. Cyber Command announced in its fiscal 2026 budget the creation of a new AI program called "Artificial Intelligence for Cyberspace Operations," designed to develop and pilot AI technologies across five categories: vulnerabilities and exploits; network security, monitoring, and visualization; modeling and predictive analytics; persona and identity; and infrastructure and transport.

This represents a shift from traditional cybersecurity approaches that relied on human analysts to detect and respond to threats. AI-powered systems can now process vast amounts of network traffic in real-time, identifying patterns and anomalies that would be impossible for human operators to detect. More concerning from a strategic perspective, these same AI capabilities can be weaponized for offensive operations, enabling what experts describe as "generative AI for cognitive electronic warfare and malware development."

The offensive potential of AI in cyber operations extends beyond traditional hacking. AI systems can now generate sophisticated phishing campaigns, create convincing fake websites, and develop polymorphic malware that adapts to defensive measures in real-time. The Pentagon's recognition that its autonomous weapons policy "does not apply to cyber" reflects the unique challenges posed by AI in the cyber domain, where the speed of operations often precludes meaningful human oversight.

Recent threat assessments highlight how adversaries are already exploiting these capabilities. The 2024 revelations about China's Volt Typhoon hacking group showed how state actors are using AI to implant malware in critical U.S. infrastructure such as water and power systems. Unlike previous cyber espionage campaigns, these operations were designed for disruption rather than intelligence gathering, representing a shift toward cyber operations intended to support potential military conflicts.

Space Applications and Integration

The space domain has become increasingly central to AI-enabled military operations, serving as both a platform for AI capabilities and a target for AI-enhanced attacks. The U.S. Space Force published its Data and Artificial Intelligence FY 2025 Strategic Action Plan, outlining how AI will be integrated across satellite operations, from autonomous collision avoidance to real-time threat detection and space situational awareness.

Current space-based AI applications span multiple mission areas. AI-enhanced satellite operations enable autonomous network management, predictive maintenance, and dynamic resource allocation across satellite constellations. The Space Force expects to launch over 100 satellites in 2025, nearly doubling its operational capacity, with many incorporating AI capabilities for autonomous operations and enhanced resilience against both kinetic and non-kinetic attacks.

Intelligence processing from space-based sensors represents another crucial application. AI systems can now analyze satellite imagery in real-time, identifying changes in ground activities, tracking mobile targets, and providing tactical intelligence to ground forces within minutes rather than hours. The Army demonstrated this capability during Project Convergence exercises, using AI to process commercial satellite data and deliver targeting information to weapon systems, reducing the time from satellite data collection to target engagement from 20 minutes to 20 seconds.

The integration of space and cyber domains through AI creates new vulnerabilities and capabilities. As General Stephen Whiting, commander of U.S. Space Command, noted, "cyberspace is the soft underbelly of our global space networks." The 2022 Russian cyberattack against Viasat, which disrupted Ukrainian communications before the ground invasion, demonstrated how cyber operations can target space systems to achieve terrestrial objectives. Modern space architectures are increasingly dependent on cybersystems for command, control, and data processing, making them vulnerable to AI-enhanced cyber attacks.

China is pursuing particularly ambitious space-based AI capabilities, beginning deployment of a network of 2,800 satellites designed to operate like terrestrial data centers in space. This represents a significant escalation in space-based computing capabilities, potentially enabling real-time AI processing of global surveillance data without the latency constraints of ground-based systems.

Current Applications Analysis

The operational deployment of AI across cyber, space, and information domains has accelerated dramatically since 2022, driven by real-world conflicts and strategic competition. Current applications span from tactical battlefield support to strategic-level influence operations, with each domain reinforcing capabilities in the others.

Information Warfare Applications

Modern information warfare operations demonstrate the sophisticated application of AI across multiple platforms and targeting methods. The Chinese approach to Taiwan exemplifies this integration, employing what intelligence analysts describe as a "three-pronged approach" characterized by disinformation, military intimidation, and influence operations.

Chinese operations use AI-powered bot swarms and paid influencers on social media to distribute low-quality but massively distributed disinformation. During Taiwan's 2024 election, content like the "Secret History of Tsai Ing-wen" was amplified across platforms using AI-generated accounts and engagement. The sophistication extended to AI-generated fake news and content insertion through cyberattacks on Taiwan's critical infrastructure, timed to coincide with electoral events.

The global scope of AI-enabled information operations became evident during the 2024 "super election year," when over 60 countries held elections affecting half the world's population. Despite predictions of widespread AI-driven electoral chaos, the impact proved more subtle but nonetheless significant. In Slovakia, AI-generated audio of opposition leader Michal Šimečka discussing election manipulation went viral days before voting, contributing to unexpected electoral outcomes. In the United States, up to 25,000 New Hampshire voters received robocalls featuring an AI-generated voice of President Biden urging them not to vote in the Democratic primary.

The effectiveness of these operations varies significantly. While Romania's election was annulled due to evidence of AI-enabled manipulation, Taiwan's Democratic Progressive Party candidate Lai Ching-te won despite extensive Chinese cognitive warfare efforts. The mixed results suggest that while AI enhances the scale and speed of information operations, their success depends on numerous factors including media literacy, institutional resilience, and counter-messaging capabilities.

Cyber Domain Operations Evolution

The integration of AI into cyber operations has transformed both offensive and defensive capabilities. Current AI applications in cybersecurity include advanced threat detection, automated incident response, and predictive analytics for identifying vulnerabilities before they can be exploited. The Pentagon's fiscal 2025 science and technology budget request of $17.2 billion allocates $1.9 billion specifically for "integrated sensing and cyber" capabilities, reflecting the priority placed on AI-enhanced cyber operations.

Defensive applications increasingly rely on machine learning algorithms to analyze network traffic patterns and identify anomalous behavior indicative of cyber attacks. These systems can process vast amounts of data in real-time, identifying threats that would overwhelm human analysts. AI-powered network defense systems can automatically isolate compromised systems, redirect traffic to avoid bottlenecks, and maintain operational capabilities during attacks.

On the offensive side, AI capabilities enable more sophisticated and persistent attacks. State actors are using AI to generate convincing phishing emails, create fake websites that closely mimic legitimate ones, and develop malware that adapts to defensive countermeasures. The Chinese Volt Typhoon group's activities demonstrate how AI can enhance traditional cyber operations, enabling the pre-positioning of malware in critical infrastructure for potential future activation.

The speed of AI-enabled cyber operations presents particular challenges for traditional response mechanisms. Cyber attacks can now unfold at machine speed, potentially overwhelming human decision-making processes. This has led to increased emphasis on automated defensive systems that can respond to threats faster than human operators, though this automation raises its own concerns about potential for escalation and loss of human control.

Space Operations Integration

Space-based applications of AI span from satellite operations to intelligence processing and threat detection. Current systems demonstrate the maturation of AI technologies for space applications, with operational satellites incorporating AI for autonomous operations, collision avoidance, and mission optimization.

The Space Force's Silent Barker satellites, which became operational in early 2025, exemplify advanced space-based AI applications. These spacecraft observe objects and activities in geosynchronous orbit, using AI to identify potentially threatening behaviors and track objects that might indicate hostile intent. The satellites represent a significant advancement in space situational awareness, providing real-time intelligence about activities in the most critical orbital regions.

Satellite constellation management increasingly relies on AI for coordinating communications, optimizing coverage, and maintaining network integrity. Modern satellite constellations like those being deployed by the Space Development Agency use AI to dynamically route communications, avoid interference, and maintain connectivity even when individual satellites are compromised or destroyed. This creates more resilient space architectures that can continue operating despite kinetic or non-kinetic attacks.

Intelligence processing from space-based sensors has been revolutionized by AI capabilities. Current systems can automatically identify changes in ground activities, track mobile targets, and provide real-time intelligence to military commanders. The integration of computer vision with large language models enables space-based systems to classify and describe observed activities in natural language, making intelligence more accessible to non-technical users.

Strategic Culture Impacts

The integration of AI across cyber, space, and information domains is fundamentally altering strategic cultures worldwide, forcing military organizations to adapt their thinking about warfare, decision-making, and operational concepts. These changes extend beyond mere technological adoption to encompass new approaches to strategy formulation and implementation.

New Operational Concepts

The emergence of AI-enabled multi-domain operations has generated entirely new strategic concepts that transcend traditional military thinking. Chinese strategists describe "intelligentized warfare" as a paradigm where AI systems "dominate information systems to effectively paralyze an opponent's joint force" through cognitive rather than kinetic means. This represents a fundamental shift from physical destruction to information and psychological dominance.

The concept of "system destruction warfare" illustrates this evolution. Rather than destroying individual platforms or facilities, AI-enabled operations target the information links between sensors and platforms, effectively neutralizing entire systems without kinetic engagement. This approach leverages AI's ability to identify and exploit critical nodes in complex networks, achieving strategic effects through precision disruption rather than mass destruction.

"Multi-domain precision warfare" represents another emerging concept, integrating AI across all operational domains to achieve coordinated effects. This approach uses AI to synchronize operations across cyber, space, electronic warfare, and kinetic domains, creating cascading effects that overwhelm enemy decision-making processes. The speed and complexity of such operations exceed human cognitive capabilities, requiring new approaches to command and control that blend human judgment with machine precision.

Cultural Adaptation Requirements

Military organizations worldwide are grappling with the cultural changes required to effectively integrate AI capabilities. The speed of AI-enabled operations creates fundamental tensions with traditional military cultures that emphasize deliberation, planning, and human oversight. Decision cycles that previously unfolded over hours or days may now occur in seconds, challenging established approaches to command and control.

The concept of "decision advantage" has emerged as a central organizing principle for AI integration. Military leaders must now think in terms of achieving superiority in decision-making speed and quality rather than simply deploying superior platforms or weapons. This requires new training approaches, organizational structures, and operational procedures that can leverage AI capabilities while maintaining human oversight of critical decisions.

Traditional approaches to escalation management face particular challenges in AI-enabled conflicts. Autonomous systems operating at machine speed may interact in unpredictable ways, potentially creating escalation dynamics that exceed human ability to control. Military cultures must therefore develop new approaches to managing AI systems that balance autonomy with human oversight, particularly in scenarios where rapid response is essential.

Attribution challenges represent another significant cultural adaptation requirement. AI-enabled operations can obscure the source of attacks, making it difficult to identify responsible parties and formulate appropriate responses. This complicates traditional deterrence models that rely on clear attribution and proportional response, requiring new approaches to conflict management in ambiguous operational environments.

Information Domain Transformation

The information domain has experienced perhaps the most dramatic transformation under AI integration. Traditional approaches to information operations focused on broadcasting messages to broad audiences through conventional media channels. AI enables personalized, targeted influence operations that can adapt content and messaging to individual users based on their psychological profiles and behavioral patterns.

Military organizations must now defend against AI-generated disinformation that can be produced at massive scale with convincing authenticity. Traditional counter-propaganda approaches that relied on authoritative sources and fact-checking are insufficient against AI systems that can generate thousands of variations of misleading content tailored to specific audiences. This requires new approaches to information defense that combine technical detection capabilities with enhanced media literacy and institutional resilience.

The speed of information operations in the AI era challenges traditional concepts of strategic communication. Information campaigns can now go viral and achieve global reach within hours, potentially shaping public opinion before authorities can formulate responses. Military organizations must therefore develop rapid response capabilities for information conflicts while maintaining accuracy and credibility in their communications.

The blurring of boundaries between domestic and international information environments presents additional challenges for democratic militaries. AI-enabled information operations can target domestic populations while ostensibly focusing on foreign audiences, raising questions about appropriate military roles in domestic information environments and the limits of democratic oversight over AI capabilities.

Alliance Implications

The integration of AI across cyber, space, and information domains creates both opportunities and challenges for alliance relationships. While AI capabilities offer the potential for enhanced cooperation and interoperability, they also introduce new complexities in information sharing, operational coordination, and strategic planning.

Coordination Challenges

NATO's experience with AI integration illustrates the challenges facing alliance structures in the AI era. The organization's revised AI strategy, released in July 2024, acknowledges that future high-speed conflicts will require "sharing large volumes of AI-parsed data to connect sensors to shooters" across alliance members. However, the strategy also warns that "computer vision systems from different nations may contradict each other and slow down targeting cycles rather than speeding them up."

The technical challenge of interoperability extends beyond simple data sharing to encompass fundamental differences in AI development approaches. Allied nations have developed AI systems using different training data, algorithms, and operational concepts, creating potential conflicts when these systems must operate together. NATO's response includes establishing "key elements of an Alliance-wide AI Testing, Evaluation, Verification & Validation landscape" through the DIANA network, but achieving true interoperability remains a work in progress.

The speed of AI-enabled operations creates particular challenges for alliance decision-making processes. Traditional NATO consultation mechanisms may be too slow for conflicts where critical decisions must be made in seconds rather than hours. This tension between the need for rapid response and the requirement for alliance consensus represents a fundamental challenge to collective defense in the AI era.

Trust and Security Dilemmas

Information sharing represents perhaps the most complex challenge facing AI-enabled alliances. The classification of AI products creates particular difficulties, as intelligence officials note that "no one's come up with great answers for appropriate classification of AI products." The phenomenon of "classification by compilation" means that AI systems can piece together classified information from supposedly unclassified sources, complicating traditional approaches to information security.

The development of national AI champions creates tensions with alliance cooperation requirements. Countries investing heavily in domestic AI capabilities may be reluctant to share advanced algorithms or training data with allies, fearing both economic competition and potential security leaks. The U.S. decision to exclude 22 NATO members from its "trusted countries" list for technology sharing illustrates these tensions, with countries like Poland and Israel protesting their exclusion despite their alliance status.

Operational security concerns extend to the protection of AI algorithms and training data during joint operations. Unlike traditional military equipment that can be physically secured, AI capabilities depend on software and data that are inherently more difficult to protect. This creates new vulnerabilities in alliance operations where AI systems must share information and coordinate activities across national boundaries.

Emerging Solutions

NATO's DIANA network represents the most ambitious effort to address alliance AI coordination challenges. The network now comprises 23 accelerator sites and 182 test centers across 28 allied countries, providing infrastructure for joint AI development and testing. The expansion of DIANA's network in 2024 doubled its capacity and demonstrated alliance commitment to cooperative AI development.

The development of common AI architectures and interfaces offers one path toward improved interoperability. Rather than attempting to harmonize existing national AI systems, allies are exploring modular approaches that enable plug-and-play integration of different AI capabilities. This approach allows countries to maintain their national AI advantages while contributing to collective capabilities.

Joint training exercises incorporating AI systems provide opportunities to test interoperability and develop common operating procedures. These exercises reveal practical challenges in AI coordination and enable the development of solutions before operational deployment. The NATO-Ukraine Innovation Cooperation Roadmap, launched at the 2024 Washington Summit, exemplifies this approach by providing a framework for testing AI capabilities in real-world conditions.

Policy frameworks for AI governance represent another area of alliance cooperation. NATO's six Principles of Responsible Use—Lawfulness, Responsibility and Accountability, Explainability and Traceability, Reliability, Governability, and Bias Mitigation—provide a common framework for AI development and deployment. Over 40 nations have joined U.S. endorsements of responsible military AI declarations, suggesting emerging consensus on AI governance principles.

Chinese Strategic Focus

China's approach to AI integration across cyber, space, and information domains reflects its broader strategic culture and represents perhaps the most systematic effort to weaponize artificial intelligence for strategic advantage. Chinese military doctrine views AI as an opportunity to leapfrog Western technological advantages and achieve strategic parity through asymmetric means.

Information Dominance Strategy

The People's Liberation Army's Strategic Support Force (SSF) has been tasked with achieving "information dominance to facilitate decision superiority" through integrated operations across space, cyber, and information domains. This mission reflects Chinese strategic thinking that views information control as the key to victory in modern conflicts, with AI serving as the primary enabler of this control.

Chinese information dominance strategy extends beyond traditional military targets to encompass civilian populations and democratic institutions. The cognitive warfare campaign against Taiwan demonstrates this approach, using AI to influence electoral outcomes and shape public opinion about unification. The deployment of "sophisticated machine learning models to create personalized propaganda aimed at Taiwan" represents a systematic effort to use AI for strategic political objectives.

The integration of domestic social control mechanisms with military information operations provides China with unique advantages in cognitive warfare. AI systems developed for domestic population monitoring and control can be adapted for external information operations, creating synergies between internal security and external strategic competition. This civil-military fusion approach enables China to field information warfare capabilities that would be difficult for democratic societies to replicate.

Technical Integration Approaches

China's military-civil fusion strategy enables rapid integration of commercial AI advances into military applications. The SSF is developing what Chinese sources describe as "intelligent satellites," "ISR software," "automated cyberattack software," and "cognitive electronic software" that integrates civilian AI research with military requirements. This approach allows China to leverage its massive commercial AI sector for military applications while maintaining plausible deniability about the military nature of civilian research.

The Chinese approach to AI integration emphasizes autonomous operations and reduced human oversight. Unlike Western approaches that maintain "human in the loop" requirements for critical decisions, Chinese doctrine appears more accepting of fully autonomous AI operations. This difference reflects broader cultural and political distinctions in approaches to human-machine interaction and may provide China with operational advantages in conflicts where speed is critical.

Quantum computing integration represents a particular area of Chinese focus, with billions invested in research aimed at "leapfrogging scientific breakthroughs" that could provide dramatic advantages in AI processing capabilities. Chinese researchers anticipate that quantum-enhanced AI could enable "increases in calculating speed that are impossible to imagine," potentially creating decisive advantages in real-time information processing and decision-making.

Global Implications

China's systematic approach to AI weaponization influences other authoritarian states and challenges Western approaches to military AI governance. The February 2024 agreement between China and Russia to "consult and coordinate on AI use in military technology" suggests emerging cooperation between authoritarian powers in AI development and deployment.

Chinese AI military applications serve as a model for other countries seeking to challenge established powers through technological means. The demonstrated effectiveness of AI-enhanced information operations against democratic societies provides a template that other authoritarian regimes can adapt to their own strategic objectives.

The scale and sophistication of Chinese AI military programs create competitive pressures that influence Western development priorities. The recognition that China "now globally dominates 57 of 64 critical technologies" in AI-related fields has prompted increased Western investment in military AI capabilities and accelerated deployment timelines for AI systems that might otherwise undergo more extensive testing and evaluation.

Emerging Challenges

The rapid integration of AI across cyber, space, and information domains creates a range of new challenges that existing governance frameworks and strategic concepts are struggling to address. These challenges span technical, operational, and strategic levels, requiring comprehensive responses from both individual nations and international organizations.

Attribution and Accountability

AI-enabled operations across multiple domains create unprecedented challenges for attribution and accountability. Traditional approaches to identifying attack sources rely on technical forensics and intelligence collection methods that may be inadequate for AI-generated content and autonomous system operations. The ability of AI systems to generate convincing false content while obscuring their origins complicates efforts to hold responsible parties accountable for their actions.

The phenomenon of "plausible deniability" has been enhanced by AI capabilities that enable state actors to conduct operations through proxies while maintaining distance from direct involvement. The use of AI-generated content allows attackers to claim that observed activities were the work of independent actors rather than state-sponsored operations, complicating diplomatic and legal responses to aggressive activities.

Cross-domain operations enabled by AI create additional attribution challenges. Attacks that begin in cyberspace but affect space systems, or information operations that influence physical world events, blur traditional boundaries between domains and legal frameworks. Existing international law and military doctrine struggle to address scenarios where AI systems operating in one domain create effects in others.

Escalation Management

The speed and complexity of AI-enabled operations create new risks for uncontrolled escalation. Autonomous systems operating across multiple domains may interact in unpredictable ways, potentially creating escalation spirals that exceed human ability to understand or control. The compression of decision timelines from hours to seconds leaves little time for human oversight or de-escalation efforts.

The integration of AI systems across domains creates potential for cascading failures that could rapidly escalate conflicts. An AI-enabled cyber attack that affects space systems could trigger autonomous defensive responses that are interpreted as offensive actions, leading to retaliation that escalates beyond the original intent of any human decision-maker.

AI systems trained on historical data may not adequately account for the novel challenges of multi-domain conflicts, potentially leading to inappropriate responses to unprecedented situations. The lack of training data for AI-enabled warfare scenarios means that AI systems may behave unpredictably when faced with complex operational environments.

Democratic Oversight and Control

The speed and complexity of AI operations challenge traditional mechanisms for democratic oversight of military activities. Autonomous systems operating at machine speed may take actions that have significant strategic implications before civilian authorities can review or approve them. This creates tension between operational effectiveness and democratic accountability.

The classification of AI capabilities and operations makes it difficult for democratic institutions to provide effective oversight. When the capabilities and limitations of AI systems are classified, elected officials may lack the information necessary to make informed decisions about their deployment and use.

International cooperation on AI governance faces particular challenges in addressing military applications. While there is growing consensus on the need for responsible AI development, translating these principles into binding international agreements remains difficult, particularly when major powers view AI capabilities as sources of strategic advantage.

Key Insight Development

The integration of artificial intelligence across cyber, space, and information domains represents more than just technological advancement—it signals the emergence of a fundamentally new form of strategic competition where traditional domain boundaries become meaningless. This transformation requires not just new technologies but entirely new approaches to strategic thinking, alliance cooperation, and international governance.

The evidence from recent conflicts and strategic competition demonstrates that AI is not merely accelerating existing forms of warfare but enabling entirely new categories of strategic action. The Chinese cognitive warfare campaign against Taiwan, Romanian election interference, and Russian information operations against Ukraine all illustrate how AI enables strategic effects that were previously impossible to achieve. These operations succeed not through superior firepower or advanced platforms but through the sophisticated manipulation of information environments and decision-making processes.

The convergence of domains through AI creates new strategic realities that existing military organizations and alliance structures are struggling to address. When a single AI system can simultaneously conduct cyber attacks, coordinate space-based surveillance, and execute information operations, traditional approaches to force structure, command and control, and operational planning become inadequate. Military organizations must develop new organizational models that can integrate AI capabilities across all domains while maintaining human oversight and democratic accountability.

Alliance relationships face particular challenges in adapting to AI-enabled strategic competition. The technical complexity of AI interoperability, combined with national sensitivities about sharing advanced capabilities, creates tensions between the benefits of cooperation and the risks of technology transfer. NATO's DIANA network and revised AI strategy represent important first steps, but achieving true alliance integration of AI capabilities will require unprecedented levels of cooperation and trust.

The speed of AI development and deployment creates a temporal mismatch between technological capabilities and governance frameworks. International law, military doctrine, and alliance structures designed for previous eras of warfare are inadequate for addressing AI-enabled strategic competition. The window for developing appropriate governance frameworks is closing rapidly as AI capabilities become more sophisticated and widely deployed.

Perhaps most significantly, the integration of AI across cyber, space, and information domains is redefining the fundamental nature of strategic advantage. Success in future conflicts may depend less on traditional measures of military capability and more on the ability to achieve cognitive dominance through superior information processing and decision-making. This shift requires strategic cultures that can blend human wisdom with machine intelligence while maintaining the values and principles that distinguish democratic societies from their authoritarian competitors.

The new high ground in strategic competition is not geographic but cognitive—the ability to shape perceptions, influence decisions, and control information flows across all domains simultaneously. Nations that master this integration while maintaining human agency and democratic oversight will shape the future international order. Those that fail to adapt risk strategic irrelevance in an era where algorithms may matter more than armies.

References

1. **Center for Strategic and International Studies**. "Securing Cyber and Space: How the United States Can Disrupt China's Blockade Plans." March 28, 2025. https://www.csis.org/analysis/securing-cyber-and-space-how-united-states-can-disrupt-chinas-blockade-plans
2. **Center for Strategic and International Studies**. "Space and Cyber Warfare as One." January 10, 2025. https://www.csis.org/analysis/space-and-cyber-warfare-one
3. **Center for Strategic and International Studies**. "2025 Space Threat Assessment: Cyberattacks on space systems persist, tracking harder amid infrastructure threats." April 28, 2025.
4. **NATO**. "NATO releases revised AI strategy." July 10, 2024. https://www.nato.int/cps/en/natohq/news_227234.htm
5. **NATO**. "DIANA, NATO's innovation accelerator, doubles the size of its transatlantic network." March 14, 2024. https://www.nato.int/cps/en/natohq/news_223648.htm
6. **NATO**. "Summary of NATO's revised Artificial Intelligence (AI) strategy." July 10, 2024. https://www.nato.int/cps/ar/natohq/official_texts_227237.htm
7. **Oxford Academic**. "How China's Cognitive Warfare Works: A Frontline Perspective of Taiwan's Anti-Disinformation Wars." Journal of Global Security Studies, July 19, 2022. https://academic.oup.com/jogss/article/7/4/ogac016/6647447
8. **The Diplomat**. "China's Chilling Cognitive Warfare Plans." May 7, 2024. https://thediplomat.com/2024/05/chinas-chilling-cognitive-warfare-plans/
9. **War on the Rocks**. "New Tech, New Concepts: China's Plans for AI and Cognitive Warfare." April 15, 2022. https://warontherocks.com/2022/04/new-tech-new-concepts-chinas-plans-for-ai-and-cognitive-warfare/
10. **Foreign Policy Research Institute**. "AI Dependence and Political Blind Spots Undermine Beijing's War Strategy." March 18, 2025. https://www.fpri.org/article/2025/03/ai-dependence-and-political-blind-spots-undermine-beijings-war-strategy/
11. **Observer Research Foundation**. "The essence of cognitive warfare: Focusing the lens toward Chinese strategies." April 8, 2024. https://www.orfonline.org/expert-speak/the-essence-of-cognitive-warfare-focusing-the-lens-toward-chinese-strategies

12. **DefenseScoop**. "Cyber Command creates new AI program in fiscal 2026 budget." July 7, 2025. https://defensescoop.com/2025/07/07/cyber-command-fy26-budget-request-new-ai-program/
13. **DefenseScoop**. "AI, space, integrated sensing and cyber dominate Pentagon's S&T funding plans." May 1, 2024. https://defensescoop.com/2024/05/01/dod-2025-budget-science-technology-ai-space-sensing-cyber/
14. **NATO Review**. "Algorithmic invasions: How information warfare threatens NATO's eastern flank." February 7, 2025. https://www.nato.int/docu/review/articles/2025/02/07/algorithmic-invasions-how-information-warfare-threatens-nato-s-eastern-flank/index.html
15. **Harvard Kennedy School Misinformation Review**. "The origin of public concerns over AI supercharging misinformation in the 2024 U.S. presidential election." February 11, 2025. https://misinforeview.hks.harvard.edu/article/the-origin-of-public-concerns-over-ai-supercharging-misinformation-in-the-2024-u-s-presidential-election/
16. **Munich Security Conference**. "AI-pocalypse Now? Disinformation, AI, and the Super Election Year." 2024. https://securityconference.org/en/publications/analyses/ai-pocalypse-disinformation-super-election-year/
17. **Carnegie Endowment for International Peace**. "Can Democracy Survive the Disruptive Power of AI?" December 2024. https://carnegieendowment.org/research/2024/12/can-democracy-survive-the-disruptive-power-of-ai
18. **Harvard Ash Center**. "The apocalypse that wasn't: AI was everywhere in 2024's elections, but deepfakes and misinformation were only part of the picture." December 4, 2024. https://ash.harvard.edu/articles/the-apocalypse-that-wasnt-ai-was-everywhere-in-2024s-elections-but-deepfakes-and-misinformation-were-only-part-of-the-picture/
19. **Space Force**. "Space Force publishes 2025 Data, AI strategic action plan." 2025. https://www.spaceforce.mil/news/article-display/article/4124351/space-force-publishes-2025-data-ai-strategic-action-plan/
20. **SpaceNews**. "Improving Space AI: Ground-to-orbit efforts aim to advance satellite intelligence." November 13, 2024. https://spacenews.com/improving-space-ai-ground-orbit-efforts-aim-advance-satellite-intelligence/
21. **Air & Space Forces Magazine**. "Space Force Will Add 100-Plus Satellites in 2025 to Boost Resilient Networks." March 21, 2025. https://www.airandspaceforces.com/space-force-100-satellites-2025-cyber-networks/
22. **Modern Diplomacy**. "The Final Frontier: Satellite Warfare and the Future of Space-Based Military Surveillance." December 23, 2024. https://moderndiplomacy.eu/2024/12/24/the-final-frontier-satellite-warfare-and-the-future-of-space-based-military-surveillance/
23. **Defense News**. "Space Force's first Silent Barker satellites to go live in early 2025."

December 11, 2024. https://www.defensenews.com/space/2024/12/11/space-forces-first-silent-barker-satellites-to-be-live-in-early-2025/

24. **Breaking Defense**. "NATO updates AI strategy for the age of ChatGPT." July 25, 2024. https://breakingdefense.com/2024/07/nato-updates-ai-strategy-for-the-age-of-chatgpt/
25. **The SAIS Review of International Affairs**. "Social Media, Disinformation, and AI: Transforming the Landscape of the 2024 U.S. Presidential Political Campaigns." January 14, 2025. https://saisreview.sais.jhu.edu/social-media-disinformation-and-ai-transforming-the-landscape-of-the-2024-u-s-presidential-political-campaigns/

Chapter 11: Economic Warfare 2.0: AI and Strategic Competition

When Silicon Valley Becomes the New Battlefield

Opening Scene: The DeepSeek Shock

On Monday, January 27, 2025, financial markets witnessed what would become known as the "DeepSeek Shock"—a single day that crystallized how artificial intelligence had transformed economic competition into a form of strategic warfare. As trading opened in New York, news of a Chinese AI startup's breakthrough sent shockwaves through global markets that would reverberate across the entire technology sector.

DeepSeek, a relatively unknown Chinese AI company founded just two years earlier in Hangzhou, had released its R1 reasoning model over the weekend. The announcement seemed routine—another AI company launching another model. But by Monday morning, the implications became clear: DeepSeek claimed to have achieved performance comparable to OpenAI's ChatGPT and GPT-4o while spending just $5.6 million on computing power, compared to the hundreds of millions or billions of dollars U.S. companies typically invested in similar capabilities.

The market reaction was swift and devastating. Nvidia, the undisputed king of AI infrastructure, saw its stock plummet 17%, erasing $588.8 billion in market value—the largest single-day loss for any company in stock market history. Other semiconductor giants followed suit: Broadcom fell 17%, ASML dropped 6%, and Micron Technology tumbled 10%. The tech-heavy Nasdaq Composite plunged 3.1%, while the broader S&P 500 declined 1.5%.

But the damage extended far beyond semiconductors. Energy companies that had positioned themselves to power the AI boom also crashed. GE Vernova, which makes wind and gas turbines, plunged 21%, while electricity generator Vistra slumped 28%. The market was pricing in a new reality: if AI could be built more efficiently, the massive infrastructure investments that companies had planned might be unnecessary.

Marc Andreessen, the influential venture capitalist and early Trump supporter, captured the moment's significance in a post on X, calling DeepSeek R1 "AI's Sputnik moment"—a reference to the 1957 satellite launch that shocked America into realizing its technological supremacy was not assured. For the first time since the AI boom began, American technological dominance appeared vulnerable.

The DeepSeek announcement came at a particularly sensitive moment. Just days earlier, President Trump had announced Stargate, a $500 billion venture with OpenAI, SoftBank, and Oracle designed to cement America's AI leadership. The timing seemed almost calculated—a reminder that in an interconnected global economy, technological leadership could shift with surprising speed.

What made the DeepSeek breakthrough particularly unsettling was how it had been achieved. Despite years of U.S. export controls designed to deny China access to advanced AI chips, DeepSeek had reportedly used "reduced capability chips from Nvidia" to train its model. The company had essentially done more with less, using software optimization and innovative architecture to overcome hardware limitations.

This was economic warfare in the 21st century—not fought with tariffs and trade barriers alone, but with algorithms and semiconductors, cloud computing and rare earth minerals. The battlefield was no longer just physical territory or traditional manufacturing, but the digital infrastructure that increasingly underpinned modern economic life.

The New Economic Battlefield

The DeepSeek incident illuminated a fundamental shift in how nations compete economically. AI had evolved from a promising technology to the central arena of strategic competition between the world's major powers. As one analyst observed, "The number-one factor that will define whether the US or China wins this race is whose technology is most broadly adopted in the rest of the world."

This transformation reflected a broader evolution in economic competition. Traditional measures of national power—military spending, manufacturing capacity, natural resources—remained important, but were increasingly being overshadowed by capabilities in artificial intelligence, semiconductor design, and digital infrastructure. The global AI market, valued at $638 billion in 2024, was projected to reach $3.68 trillion by 2034, with a compound annual growth rate of 19.2%. Control over this market would determine which nations shaped the economic and technological landscape of the coming decades.

The economic implications extended far beyond the technology sector itself. AI was becoming the critical enabler for everything from manufacturing automation to financial services, from healthcare delivery to military systems. Nations that controlled AI development would gain advantages across their entire economies, while those dependent on foreign AI technology would find themselves in positions of strategic vulnerability.

The Infrastructure Arms Race

The scale of investment required for AI leadership had reached staggering levels. American technology giants alone were planning to spend over $320 billion on AI technologies and data center buildouts in 2025, up from $230 billion in 2024. Amazon led with plans to spend over $100 billion, while Microsoft, Google, and Meta each committed tens of billions more.

This infrastructure arms race reflected the reality that AI capabilities required massive computational resources. Training advanced AI models demanded access to thousands of high-end graphics processing units (GPUs), specialized data centers, and enormous amounts of electricity. The companies and countries that could marshal these resources would determine the direction of AI development.

But the DeepSeek announcement suggested that this conventional wisdom might be flawed. If efficiency gains could deliver comparable results at a fraction of the cost, the entire foundation of the AI economy might shift. As one market analyst noted, "DeepSeek has taken the market by storm by doing more with less. They activate only the most relevant portions of their model for each query, and that saves money and computation power."

Economic Policy as National Security Strategy

The blurring of economic and national security policy had become one of the defining characteristics of AI competition. Unlike previous technological revolutions, AI development was viewed by major powers not just as an economic opportunity, but as a matter of national survival.

This perspective was evident in how governments approached AI regulation and investment. The United States had implemented comprehensive export controls on advanced semiconductors to China, while investing over $50 billion through the CHIPS and Science Act to rebuild domestic semiconductor manufacturing capacity. China had responded with its own massive investments in AI development and restrictions on critical mineral exports.

The European Union had taken a different approach, emphasizing regulatory leadership through the AI Act while launching initiatives like the €1 billion NATO Innovation Fund and AI factories across the continent. Each approach reflected different theories about how to achieve and maintain competitive advantage in the AI era.

Supply Chain Weaponization: The New Economic Arsenal

The integration of AI into the global economy had created new vulnerabilities that nations were learning to exploit as weapons of economic warfare. Unlike traditional trade disputes focused on finished goods, AI competition involved complex supply chains spanning multiple countries and industries, creating numerous pressure points for strategic manipulation.

The Semiconductor Chokehold

The most visible example of supply chain weaponization involved semiconductors—the specialized chips essential for AI development. The United States had leveraged its dominance in semiconductor design software and manufacturing equipment to restrict Chinese access to advanced AI chips. These controls, implemented in October 2022 and tightened repeatedly since, targeted "chokepoint" technologies where American companies held decisive advantages.

The strategy was comprehensive. By restricting access to electronic design automation (EDA) software—dominated by three American companies (Mentor Graphics, Cadence Design Systems, and Synopsys)—the United States made it nearly impossible for Chinese companies to design advanced AI chips. Simultaneously, export controls on semiconductor manufacturing equipment from companies like ASML in the Netherlands prevented Chinese foundries from producing cutting-edge processors.

The Foreign Direct Product Rule extended these controls globally, prohibiting any semiconductor manufacturing company worldwide from providing services to Chinese chip design companies seeking to make high-end chips for AI applications. As one analysis noted, "Any chip manufacturing operation—whether Chinese or otherwise—that seeks to build Chinese chip designs will risk losing its own access to U.S. semiconductor manufacturing equipment."

China's Mineral Counter-Offensive

China's response demonstrated how economic warfare in the AI era involved multiple layers of supply chain vulnerability. In December 2024, following another round of U.S. semiconductor restrictions, China imposed export controls on gallium, germanium, antimony, and other "superhard materials" essential for semiconductor production.

This retaliation highlighted China's own strategic advantages. Despite U.S. dominance in semiconductor design and manufacturing equipment, China controlled the mining and processing of rare earth elements essential for chip production. China produced approximately 60% of the world's rare earths and controlled over 90% of rare earth processing capacity.

The materials targeted by China's export restrictions were carefully chosen for maximum impact:

- **Gallium and Germanium**: Critical for semiconductor production, with China controlling approximately 80% of global gallium production and 60% of germanium production
- **Antimony**: Essential for military applications including ammunition, infrared missiles, and night-vision technology, with China producing 48% of global supply
- **Graphite**: Vital for lithium-ion battery production and electric vehicle manufacturing, with China dominating 78% of natural graphite production

The strategic implications were profound. As one defense analyst noted, "China is acquiring advanced weapons systems and equipment five to six times faster than the U.S. Further bans on critical minerals inputs will only widen the gap, enabling China to strengthen its military capabilities more quickly than the United States."

The Vulnerability of Allied Supply Chains

The complexity of modern supply chains created vulnerabilities that extended beyond the primary U.S.-China competition. Taiwan, despite its close security relationship with the United States, produced over 60% of the world's semiconductors and over 90% of the most advanced chips. This concentration created what experts called a "single point of failure" for the global economy.

Bloomberg Economics estimated that a conflict over Taiwan would cost the global economy $10 trillion in the first year—a figure that, even if exaggerated, indicated the catastrophic economic consequences of supply chain disruption. European leaders recognized similar vulnerabilities, with one EU official noting that "all of the leading-edge chips that are critical to our infrastructure, to AI, to our national security ambitions, automotive manufacturing even, are being built in one part—a fragile part of the world."

The challenge for middle powers was navigating between competing economic blocs without sacrificing their own technological development. As supply chains became weaponized, countries faced pressure to choose sides in what increasingly resembled a technological Cold War.

Digital Sovereignty Strategies: National Responses to AI Dependence

The recognition that AI capabilities had become essential for national competitiveness led governments worldwide to develop strategies for "digital sovereignty"—the ability to control their own technological destiny rather than depend on foreign powers. These strategies reflected different national cultures, political systems, and economic capabilities, but shared a common goal of reducing strategic vulnerabilities.

The American Model: Innovation Through Competition

The United States approach combined massive public investment with a relatively hands-off regulatory environment designed to preserve the dynamism of private sector innovation. The CHIPS and Science Act represented the largest federal investment in technology since the space program, providing over $50 billion to rebuild domestic semiconductor manufacturing capacity.

American strategy relied heavily on leveraging the country's advantages in innovation and capital markets. U.S. venture capital firms provided the majority of global AI funding, while American universities and research institutions produced much of the fundamental research underlying AI development. Major technology companies like Microsoft, Google, Amazon, and OpenAI maintained leadership in AI model development and deployment.

The challenge for American policymakers was maintaining these advantages while protecting them from foreign exploitation. New rules required cloud computing providers to keep "50% of compute power inside the US" while capping international installations. The AI Diffusion Framework, announced in January 2025, aimed to prevent the transfer of advanced AI capabilities to adversaries while maintaining America's position as the global center of AI innovation.

The Chinese Model: State-Directed Development

China's approach reflected its political system's capacity for coordinated, long-term planning and massive resource mobilization. The Chinese government launched a 300 billion yuan ($41 billion) fund for semiconductor industry investments in 2023, supplementing 200 billion yuan ($27 billion) invested in 2019. These sums dwarfed comparable investments by Western governments.

Chinese strategy emphasized developing domestic alternatives to Western technologies across the entire AI supply chain. The country had created indigenous deep learning frameworks including PaddlePaddle (Baidu) and MindSpore (Huawei), while developing "Gitee as a domestic alternative to GitHub." The Belt and Road Initiative had evolved to emphasize "technology transfers and integration" rather than traditional infrastructure development.

The "Made in China 2025" guidance established ambitious targets, including 80% localization of Chinese chip consumption by 2030, up from approximately 24% in 2020. While many observers viewed these targets as aspirational, the commitment of resources suggested serious intent to achieve technological independence from Western suppliers.

The European Model: Regulatory Leadership

The European Union pursued a distinctive strategy emphasizing regulatory frameworks and ethical standards rather than attempting to match American or Chinese capabilities in raw computing power or state investment. The AI Act, which became law in 2024, established the world's first comprehensive regulatory framework for artificial intelligence development and deployment.

EU leaders argued that regulatory leadership could become a form of digital sovereignty. As French President Emmanuel Macron observed, "The future of AI is a political stake, of sovereignty and strategic dependence. Whoever sets the AI rules exports their worldview." The "Brussels Effect"—the EU's ability to set global standards through its large market size—had previously shaped everything from data protection to antitrust policy.

The EU's approach emphasized "jurisdictional independence" rather than competitive advantage, focusing on ensuring that AI development within Europe conformed to European values and legal frameworks. The €1 billion NATO Innovation Fund and AI factories across the continent represented attempts to build European AI capabilities, while the Digital Markets Act and Digital Services Act created frameworks for governing the use of AI technologies.

However, critics argued that European regulatory leadership came at the cost of innovation and competitiveness. A February 2025 Bertelsmann Stiftung report estimated that achieving complete digital sovereignty would require €300 billion ($341 billion) over the next decade, highlighting the enormous challenge facing European policymakers.

Techno-Nationalism and the Fragmentation of the Global Economy

The competition for AI dominance was driving a broader trend toward techno-nationalism—the belief that technological capabilities were too important to be left to market forces alone. This shift was fragmenting the global economy into competing technological blocs, each with its own standards, supply chains, and spheres of influence.

The End of Technology Globalization

For decades, the technology industry had exemplified economic globalization, with complex supply chains spanning multiple continents and free flow of talent, capital, and ideas. The AI revolution was disrupting this model, as governments increasingly viewed technological dependence as a national security vulnerability.

The process had accelerated dramatically since 2020. What began as targeted restrictions on specific Chinese companies had evolved into comprehensive efforts to reshape global technology supply chains. The United States had restricted 22 NATO members from its "trusted countries" list for AI technology sharing, creating tensions even among close allies.

European leaders expressed growing frustration with American restrictions that treated "disparate countries such as India and Tuvalu the same," while ally countries like Poland and Israel protested their exclusion from trusted technology partnerships. These disputes highlighted how AI competition was straining traditional alliance relationships.

Standards Wars and the Battle for Global Influence

The fragmentation of the global economy was creating competing technological ecosystems with different standards, protocols, and governance frameworks. This "standards war" would determine which countries and companies could shape the global development of AI technologies.

China's success with DeepSeek demonstrated how technological innovation could occur outside Western-dominated ecosystems. Chinese AI companies were developing capabilities that challenged American assumptions about technological leadership, while Chinese technical standards were gaining adoption in countries participating in the Belt and Road Initiative.

The United States was fighting back through initiatives like the Chip 4 alliance with Japan, South Korea, and Taiwan, aimed at securing semiconductor supply chains and limiting China's technological advancement. American officials argued that preserving democratic values in technology development required collective action among like-minded countries.

Europe's attempt to lead through regulation represented a third model, emphasizing governance frameworks rather than raw technological capabilities. The success of this approach would depend on whether other countries chose to adopt European standards or were drawn to American innovation or Chinese efficiency.

The Economics of Technological Decoupling

The costs of technological fragmentation were becoming increasingly apparent. Companies faced growing pressure to choose between markets, often requiring them to develop separate product lines for different regions. The efficiency gains from global supply chain optimization were being sacrificed for the security benefits of technological independence.

For smaller countries, the pressure to choose sides was particularly acute. As one analyst observed, "Presented with a binary choice between US and Chinese technology, middle powers must balance realism and ambition in maintaining sovereignty over their technology." The risk was that countries without the resources to develop independent capabilities would become technological dependencies of larger powers.

The emergence of "tech sovereignty" strategies reflected recognition that complete technological independence was impossible for most countries, but that strategic autonomy in key areas was essential. India's development of sovereign digital ID and payment systems demonstrated how middle powers could carve out independent capabilities in specific domains.

Strategic Implications: Economic Security as National Security

The transformation of AI into an arena of strategic competition had profound implications for how nations understood and pursued security in the 21st century. Traditional distinctions between economic policy and national security were dissolving, as technological capabilities became inseparable from military power and geopolitical influence.

The Military-Economic Nexus

The integration of AI into military systems meant that economic competition directly influenced military capabilities. Nations that led in AI development would possess decisive advantages in everything from missile defense to cyber warfare to intelligence analysis. Conversely, countries dependent on foreign AI technologies would face vulnerabilities in their most sensitive security systems.

This dynamic was evident in how military procurement was evolving. In 2024, OpenAI revised its usage guidelines to allow advanced AI systems to be deployed for weapons development and warfare applications, while Google ended a long-standing ban on using AI for military and surveillance tools. The boundaries between civilian and military AI applications were becoming increasingly blurred.

Defense contractors were struggling to keep pace with rapid innovation in the commercial AI sector. Traditional defense firms that operated on decade-long development cycles found themselves competing with technology startups that could develop and deploy new capabilities in months. This shift was forcing fundamental changes in how governments approached military procurement and industrial policy.

Alliance Relationships in the AI Era

AI competition was reshaping alliance relationships, creating new forms of interdependence while straining traditional partnerships. Countries that had been reliable allies in military and diplomatic affairs found themselves on opposite sides of technology trade disputes.

The challenge was particularly acute for NATO, where alliance members needed to coordinate AI capabilities while maintaining competitive advantages in their domestic technology industries. As one alliance official noted, "Future high-speed conflicts require sharing large volumes of AI-parsed data to connect sensors to shooters," but this coordination risked revealing sensitive technological capabilities.

The development of common standards for AI systems across allied militaries represented both an opportunity and a challenge. Successfully coordinated AI systems could provide unprecedented advantages in joint operations, but the development process required levels of trust and technological sharing that strained traditional security protocols.

Economic Deterrence and Strategic Stability

The weaponization of economic relationships through AI and technology restrictions created new forms of deterrence that operated outside traditional military frameworks. Countries could inflict significant damage on adversaries through technology export controls, supply chain disruptions, and restrictions on critical materials.

This economic deterrence operated on different timescales and through different mechanisms than military deterrence. While military conflicts involved immediate and visible escalation, economic warfare could unfold gradually through seemingly technical decisions about standards, procurement, and regulatory frameworks.

The challenge for policymakers was managing these new forms of competition without triggering economic crises that could destabilize the global economy. The DeepSeek incident demonstrated how quickly market confidence could shift based on perceptions of technological leadership, creating systemic risks that extended far beyond the technology sector.

Private Sector Integration: When Corporations Become Strategic Assets

The central role of private companies in AI development created new challenges for governments seeking to maintain strategic control over critical technologies. Unlike previous periods when military and civilian technologies developed separately, AI capabilities were primarily emerging from commercial research and development, forcing governments to find new ways to harness private sector innovation for national purposes.

The New Military-Industrial-Digital Complex

The emergence of what some analysts called the "military-industrial-digital complex" reflected the growing importance of technology companies in national security. Major AI developers like OpenAI, Google, Microsoft, and Anthropic were becoming strategic assets comparable to traditional defense contractors, but with business models and corporate cultures that operated according to different principles.

This shift created new tensions between corporate interests and national security requirements. Technology companies that had built global businesses were being asked to restrict their operations to serve national policy goals. The result was a complex dance between government agencies seeking to harness corporate capabilities and companies trying to maintain global market access.

The partnership between OpenAI and the Pentagon, announced as part of the Stargate initiative, exemplified these new relationships. The deal provided the government with access to cutting-edge AI capabilities while giving the company resources for continued development. However, critics worried about the concentration of critical capabilities in a small number of private entities.

Venture Capital as Strategic Infrastructure

The role of venture capital in AI development had elevated private investment firms to positions of strategic importance. American venture capital firms provided the majority of global AI funding, giving them influence over which technologies developed and which companies succeeded. This private sector role in determining technological trajectories created new forms of strategic competition.

Chinese government investment strategies attempted to replicate the American venture capital model while maintaining state direction over technological development. European initiatives like the NATO Innovation Fund represented attempts to create alternative funding mechanisms that could compete with American and Chinese approaches.

The challenge for governments was maintaining influence over AI development without stifling the innovation that made their technology sectors competitive. Heavy-handed government intervention risked driving investment and talent to countries with more favorable environments, but laissez-faire approaches risked losing control over strategically critical technologies.

Market Dynamics and Strategic Competition

The structure of AI markets created natural tendencies toward concentration and winner-take-all dynamics that had strategic implications. Training advanced AI models required enormous computational resources, creating barriers to entry that limited the number of companies capable of frontier AI development. The "big five" American companies—Microsoft, Google, Amazon, Meta, and OpenAI—dominated global AI capabilities, while Chinese companies like Baidu, Alibaba, and ByteDance controlled domestic markets.

This concentration created both opportunities and vulnerabilities for governments. Countries with dominant AI companies possessed significant leverage over global markets, but also faced risks if those companies were disrupted by foreign competition or technological breakthroughs. The DeepSeek incident demonstrated how quickly market leadership could shift in rapidly evolving technology sectors.

The challenge was particularly acute for smaller countries that lacked domestic AI champions. These nations faced choices between supporting local technology development, partnering with foreign companies, or accepting technological dependence. Each approach involved different trade-offs between innovation, security, and sovereignty.

Global Supply Chain Restructuring: From Efficiency to Resilience

The weaponization of supply chains in AI competition was forcing a fundamental restructuring of global economic relationships. The efficiency-focused "just-in-time" supply chain management that had characterized globalization was giving way to "just-in-case" approaches that prioritized resilience and redundancy over cost optimization.

Friend-Shoring and Near-Shoring Strategies

Governments and companies were implementing "friend-shoring" strategies that prioritized trade relationships with politically aligned countries, even at the cost of economic efficiency. The United States was encouraging semiconductor production in allied countries like Japan, South Korea, and

Taiwan while restricting technology transfers to China. Similar dynamics were emerging in other critical technology sectors.

European initiatives emphasized "near-shoring" to reduce dependence on both American and Chinese suppliers. The EU's Chips Act aimed to increase European semiconductor production capacity, while various programs supported the development of European alternatives to foreign cloud computing and AI platforms.

These strategies involved significant economic costs. Rebuilding supply chains required massive investments in new production facilities, often in countries with higher labor and regulatory costs than existing suppliers. However, governments increasingly viewed these costs as necessary insurance against supply chain disruption.

The Economics of Strategic Redundancy

The shift toward supply chain resilience was changing the fundamental economics of technology production. Instead of optimizing for single suppliers in the most cost-effective locations, companies were building redundant capabilities across multiple geographic regions and political jurisdictions.

This redundancy extended to research and development activities. Major technology companies were establishing research facilities in multiple countries to ensure continued access to global talent pools and to hedge against political restrictions on technology transfer. The result was a more distributed but less efficient global innovation ecosystem.

The costs of this transition were substantial. Industry estimates suggested that complete technological decoupling between the United States and China could reduce global economic output by trillions of dollars over the coming decade. However, policymakers increasingly viewed these costs as acceptable given the strategic risks of continued technological interdependence.

Critical Materials and Resource Security

The control of rare earth minerals and other critical materials had become a central element of AI economic warfare. China's dominance in rare earth processing gave it significant leverage over technology supply chains, while American control of semiconductor design tools provided reciprocal influence.

The challenge for countries seeking technological independence was developing alternative sources for critical materials. The United States was investing in domestic rare earth processing capabilities, with companies like MP Materials receiving government support to develop integrated supply chains from mining to magnet production. However, these efforts faced significant technical and economic obstacles.

The timeline for developing alternative supply chains meant that countries would remain vulnerable to supply disruptions for years or decades. As one analysis noted, "Developing mining and processing capabilities requires a long-term effort, meaning the United States will be on the back foot for the foreseeable future."

Future Trajectory: The Emergence of Economic Blocs

The competition for AI dominance was creating new forms of economic organization that transcended traditional trade relationships. Instead of a single global economy governed by multilateral institutions, the world was evolving toward multiple economic blocs organized around technological capabilities and geopolitical alignments.

The American Innovation Ecosystem

The United States was leveraging its advantages in venture capital, higher education, and corporate research to maintain leadership in AI innovation. American universities produced much of the fundamental research underlying AI development, while Silicon Valley's unique culture of risk-taking and rapid iteration continued to generate breakthrough technologies.

However, maintaining this ecosystem required careful balance between openness and security. Restrictions on Chinese students and researchers risked reducing the talent pool that had historically driven American innovation, while export controls could limit the global markets that made American companies competitive.

The challenge for American policymakers was preserving the conditions that made Silicon Valley successful while protecting the technologies it produced from foreign exploitation. This balancing act would determine whether the United States could maintain its technological leadership in an increasingly competitive global environment.

The Chinese Development Model

China's state-directed approach to AI development was producing impressive results, as demonstrated by companies like DeepSeek. The combination of government investment, protected domestic markets, and access to vast amounts of data created unique advantages for Chinese AI companies.

The Chinese model emphasized long-term planning and coordinated development across multiple sectors. Government agencies worked with private companies to identify strategic priorities and mobilize resources for breakthrough technologies. This approach allowed China to make rapid progress in areas targeted for development, even when facing restrictions on access to foreign technologies.

However, the sustainability of this model remained uncertain. State direction could accelerate development in targeted areas but might limit the kind of fundamental innovation that had historically driven technological breakthroughs. The challenge for Chinese policymakers was maintaining innovation while preserving political control over technological development.

The European Regulatory Framework

Europe's emphasis on regulatory leadership represented a distinctive approach to AI competition that prioritized governance frameworks over raw technological capabilities. The AI Act and related legislation created comprehensive frameworks for AI development and deployment that could influence global standards through the Brussels Effect.

This approach reflected European strengths in multilateral cooperation and institutional development, as well as recognition that Europe lacked the resources to compete directly with American or Chinese AI investments. By setting high standards for AI development, European leaders hoped to shape global technology development according to European values.

The success of this strategy would depend on whether other countries chose to adopt European standards or were drawn to the innovation capabilities of American companies or the efficiency of Chinese alternatives. The test case would be how global markets responded to the different approaches represented by American innovation, Chinese efficiency, and European governance.

Conclusion: The New Reality of Economic Warfare

The transformation of economic competition through artificial intelligence represented a fundamental shift in how nations pursued power and influence in the 21st century. The DeepSeek shock that opened this chapter was not an isolated incident but a harbinger of a new era in which technological capabilities determined national strength and economic success.

The weaponization of supply chains, the emergence of techno-nationalism, and the fragmentation of global markets created new forms of strategic competition that operated according to different rules than traditional military or economic rivalry. In this new environment, control over algorithms, semiconductors, and data infrastructure became as important as control over territory or natural resources.

The challenge for policymakers was managing this transformation without triggering economic crises that could destabilize the global system. The interconnected nature of modern economies meant that technological decoupling carried enormous costs, but the strategic risks of technological dependence were increasingly viewed as unacceptable.

The countries and companies that successfully navigated this transition would shape the economic and technological landscape of the coming decades. Those that failed to adapt risked finding themselves relegated to positions of permanent disadvantage in an increasingly AI-driven global economy.

As the competition intensified, the stakes continued to rise. AI was not just another technology sector but the foundation upon which future economic growth, military capability, and geopolitical influence would be built. The battle for AI supremacy had become the defining strategic competition of the 21st century, with consequences that would reverberate through every aspect of international relations.

The question was no longer whether economic competition would be transformed by artificial intelligence, but how quickly nations could adapt their strategies to succeed in this new reality. The answer would determine the shape of the global economy for generations to come.

Sources and References

1. CBS News. "What is DeepSeek, and why is it causing Nvidia and other stocks to slump?" https://www.cbsnews.com/news/what-is-deepseek-ai-china-stock-nvidia-nvda-asml/
2. CNN Business. "A shocking Chinese AI advancement called DeepSeek is sending US stocks plunging." January 27, 2025. https://www.cnn.com/2025/01/27/tech/deepseek-stocks-ai-china
3. NBC News. "What we know on DeepSeek AI: How it rivals ChatGPT, stock market impact, security concerns and more." January 28, 2025. https://www.nbcnews.com/tech/tech-news/deepseek-ai-model-china-chat-gpt-tech-stocks-what-know-rcna189604
4. Foreign Policy Research Institute. "Breaking the Circuit: US-China Semiconductor Controls." September 16, 2024. https://www.fpri.org/article/2024/09/breaking-the-circuit-us-china-semiconductor-controls/
5. Center for Strategic and International Studies. "Understanding U.S. Allies' Current Legal Authority to Implement AI and Semiconductor Export Controls." April 2, 2025. https://www.csis.org/analysis/understanding-us-allies-current-legal-authority-implement-ai-and-semiconductor-export
6. Center for Strategic and International Studies. "Choking off China's Access to the Future of AI." October 15, 2024. https://www.csis.org/analysis/choking-chinas-access-future-ai
7. Center for Strategic and International Studies. "The Consequences of China's New Rare Earths Export Restrictions." April 16, 2025. https://www.csis.org/analysis/consequences-chinas-new-rare-earths-export-restrictions
8. CNBC. "U.S.' inability to replace rare earths supply from China poses a threat to its defense, warns CSIS." April 15, 2025. https://www.cnbc.com/2025/04/15/us-is-unable-to-replace-rare-earths-supply-from-china-warns-csis-.html
9. Time. "Rare Earth Elements Are a Key Weapon in US-China Trade War." April 9, 2025. https://time.com/7275818/rare-earth-china-tariffs-metals-minerals/
10. Chatham House. "China's rare earth export restrictions threaten Washington's military primacy." April 14, 2025. https://www.chathamhouse.org/2025/04/chinas-rare-earth-export-restrictions-threaten-washingtons-military-primacy
11. Carnegie Endowment for International Peace. "The EU's AI Power Play: Between Deregulation and Innovation." https://carnegieendowment.org/research/2025/05/the-eus-ai-power-play-between-deregulation-and-innovation?lang=en
12. Chatham House. "The US–China AI race is forcing countries to reconsider who owns their digital infrastructure." May 13, 2025. https://www.chathamhouse.org/2025/05/us-china-ai-race-forcing-countries-reconsider-who-owns-their-digital-infrastructure

13. CNBC. "Tech megacaps plan to spend more than $300 billion in 2025 as AI race intensifies." February 8, 2025. https://www.cnbc.com/2025/02/08/tech-megacaps-to-spend-more-than-300-billion-in-2025-to-win-in-ai.html
14. Sequoia Capital. "AI in 2025: Building Blocks Firmly in Place." December 9, 2024. https://www.sequoiacap.com/article/ai-in-2025/
15. Precedence Research. "Artificial Intelligence (AI) Market Size to Hit USD 3,680.47 Bn by 2034." June 2, 2025. https://www.precedenceresearch.com/artificial-intelligence-market
16. Atlantic Council. "Why tariffs on AI hardware could undermine US competitiveness." June 15, 2025. https://www.atlanticcouncil.org/blogs/new-atlanticist/why-tariffs-on-ai-hardware-could-undermine-us-competitiveness/
17. Hudson Institute. "AI, National Security, and the Global Technology Race: How US Export Controls Define the Future of Innovation." https://www.hudson.org/national-security-defense/ai-national-security-global-technology-race-how-us-export-controls-define-nury-turkel
18. Stanford Cyber Policy Program. "Taking Stock of the DeepSeek Shock." February 5, 2025. https://cyber.fsi.stanford.edu/publication/taking-stock-deepseek-shock

Chapter 12: Alliance Politics in the AI Era: Trust and Technology

Sharing Algorithms While Maintaining Advantages

Opening Scene: The NATO AI Coordination Challenge

At NATO Headquarters in Brussels, alliance officials gathered in July 2024 faced a fundamental challenge that would have been inconceivable just a decade earlier. As they released NATO's revised artificial intelligence strategy, they confronted a reality where "future high-speed conflicts require sharing large volumes of AI-parsed data to connect sensors to shooters," yet where the very systems meant to accelerate decision-making could potentially "contradict each other and slow down targeting cycle rather than speeding it up" if not properly coordinated between allies.

The challenge was stark: how could 32 sovereign nations, each developing their own AI military capabilities, ensure these systems could work together when every second might matter in combat? The answer would reshape the fundamental nature of alliance relationships in the 21st century.

The Interoperability Challenge: When Algorithms Must Speak the Same Language

Technical Coordination Problems

The heart of NATO's AI challenge lies in a deceptively simple requirement: interoperability. As the alliance's 2024 revised AI strategy acknowledges, "Growing interoperability between all AI systems throughout the Alliance" represents one of the most critical objectives for collective defense in the AI age.

The technical challenges are profound. Computer vision systems developed by different nations using different training data and algorithms could potentially "contradict each other and slow down targeting cycle rather than speeding it up," according to NATO analysis. Unlike traditional military equipment where standardization focuses on physical compatibility—ensuring ammunition fits weapons or communication systems operate on the same frequencies—AI interoperability requires harmonizing the decision-making processes of intelligent systems themselves.

Autonomous systems must be "programmed to operate together without relying on common sense or relationships between human operators," a fundamental shift from human-led coordination that has characterized alliance operations for decades. This means developing common architectural standards, shared training protocols, and synchronized decision algorithms across multiple national defense establishments.

Speed vs. Coordination Dilemma

The promise of AI-enhanced military operations—decision cycles "measured in seconds, not days"—creates unprecedented challenges for alliance coordination. Traditional NATO procedures, developed for human-paced decision-making, become bottlenecks when AI systems can process information and recommend actions in milliseconds.

U.S. Army Undersecretary Gabe Camarillo has emphasized that "if AI is going to be used to support NATO decisionmaking, then ensuring relevant data can be successfully accessed and shared when and where it is needed will be fundamental." Without such capabilities, decisions could be made on "old or irrelevant data and coordination between allies could be inhibited, significantly limiting some of the perceived advantages of AI-enabled decision support systems."

The alliance faces the paradox of needing to maintain collective decision-making while enabling the speed advantages that make AI valuable in the first place. This tension appears in everything from target selection processes to resource allocation decisions.

Current Alliance AI Initiatives: Building the Foundation

NATO's Institutional Framework Evolution

NATO's approach to managing AI cooperation has evolved rapidly since 2021. The alliance's Data and Artificial Intelligence Review Board (DARB), established in October 2022, serves as the central coordinating body for AI standards and practices across the alliance. The DARB "must ensure that its outputs do not raise unnecessary barriers to adopting AI at the speed of relevance" while developing "practical RAI toolkits" and "supporting Allies in their RAI efforts."

The board's mandate includes developing certification standards that can ensure AI systems across the alliance meet common safety and reliability criteria. In February 2023, NATO began developing "a user-friendly and responsible Artificial Intelligence (AI) certification standard to help industries and institutions across the Alliance make sure that new AI and data projects are in line with international law, as well as NATO's norms and values."

DIANA and the NATO Innovation Fund: Accelerating Cooperation

The Defence Innovation Accelerator for the North Atlantic (DIANA) represents NATO's most ambitious attempt to foster transatlantic cooperation on critical technologies. Launched in 2022, DIANA has rapidly expanded its network to comprise "23 accelerator sites (up from 11) and 182 test centres (up from 90) in 28 Allied countries," significantly augmenting the alliance's capacity to support AI innovation.

DIANA's approach focuses on practical cooperation through challenge-based innovation. In its first year, the program received "over 1,300 applications from innovators across the Alliance" and selected "44 companies from 19 countries" for its inaugural cohort, addressing challenges in energy resilience, undersea sensing and surveillance, and secure information sharing.

The program is complemented by the NATO Innovation Fund, "the world's first multi-sovereign venture capital fund," which will "invest 1 billion euros in start-ups developing technologies for defence and security" over a 15-year period. The fund's first investments in June 2024 included companies developing solutions in "novel materials and manufacturing, AI and robotics," demonstrating the alliance's commitment to fostering innovation across critical technology areas.

Testing and Certification Landscape

NATO's revised 2024 AI strategy calls for establishing "key elements of an Alliance-wide AI Testing, Evaluation, Verification & Validation (TEV&V) landscape able to support the adoption of responsible AI." This represents a significant expansion from the 2021 strategy's more tentative language about potential cooperation.

The strategy emphasizes that "Allied testing facilities need to be able to determine if AI applications can be used safely and in accordance with NATO's PRUs [Principles of Responsible Use]." This creates a practical framework for ensuring AI systems can meet alliance standards before deployment in collaborative operations.

Trust and Security Dilemmas: The Classification Challenge

Information Sharing in the AI Age

One of the most complex challenges facing NATO AI cooperation involves managing classified information in systems that constantly generate new analyses and insights. As one NATO official acknowledged, "no one's come up with great answers for appropriate classification of AI products," particularly given the phenomenon of "classification by compilation" where "AI can piece together classified information from unclassified sources."

This challenge becomes particularly acute when AI systems are "constantly creating new analyses" that may reveal sensitive information or methods, potentially outpacing existing security review processes. Traditional classification systems, designed for discrete documents and reports, struggle to adapt to AI systems that continuously process and reprocess information.

The problem extends to sharing AI training data and algorithms themselves. Different allies may have varying approaches to data classification, creating obstacles to developing truly interoperable systems that require shared training datasets or algorithm transparency.

Technology Transfer and Competitive Dynamics

The tension between sharing AI capabilities and maintaining competitive advantages creates ongoing friction within the alliance. Recent U.S. export controls demonstrate this challenge: new regulations announced in January 2025 place "restrictions on advanced AI chips, cloud access, and model weights," creating different tiers of technology access even among allies.

The regulations divide countries into three tiers, with only "18 close U.S. allies" receiving unlimited access to advanced AI technologies, while other NATO members like Poland find themselves subject to quotas that limit their ability to import more than "1,700 high-end GPUs each year without a license from the U.S. government." Poland has specifically protested its exclusion from the "trusted countries" list, highlighting how technology restrictions can strain alliance relationships.

Burden-Sharing Evolution: AI as a New Metric

Beyond Defense Spending: Technological Capabilities

Traditional NATO burden-sharing discussions have focused on the alliance's goal that members spend 2% of GDP on defense. However, the AI age is forcing a reconsideration of how to measure meaningful contributions to collective security. Canada's 2024 Defence AI Strategy exemplifies this challenge, as the country seeks to "remain credible and interoperable" while developing specialized AI capabilities that may not appear in traditional defense spending metrics.

The challenge for smaller allies is particularly acute. They must balance developing sufficient AI capabilities to remain "credible" partners while avoiding duplicative investments that waste limited resources. This is driving toward specialization strategies where different allies lead in different AI domains rather than attempting comprehensive capabilities development.

NATO's Innovation and Burden-Sharing Integration

The NATO Innovation Fund represents a new model of burden-sharing where countries contribute to collective technological advancement rather than just traditional military capabilities. With "23 NATO Allies as Limited Partners" (becoming 24 when Sweden joined), the fund demonstrates how allies can pool resources for emerging technologies while maintaining national sovereign decision-making.

This approach acknowledges that in the AI age, innovation capacity and technological infrastructure may be as important as traditional military hardware for collective defense. Countries like Estonia and Denmark, which might struggle to meet traditional burden-sharing expectations, can make outsized contributions through digital expertise and cyber defense capabilities.

Emerging Solutions: Frameworks for Cooperation

Technical Standards and Certification

NATO is developing comprehensive technical standards for AI interoperability through its DARB framework. The alliance's six Principles of Responsible Use—"Lawfulness, Responsibility and Accountability, Explainability and Traceability, Reliability, Governability and Bias Mitigation"—provide the foundation for common standards that can enable interoperability while maintaining ethical guidelines.

The alliance is working to establish "standards, assessment templates, review processes, and other tools and good practices to operationalise the responsible adoption of AI across the Alliance." This includes developing "RAI certification standards" that can verify AI systems meet alliance requirements before integration into collaborative operations.

The Global Responsible AI Declaration

Beyond NATO's internal frameworks, nearly 60 nations now endorse the "Political Declaration on Responsible Military Use of Artificial Intelligence and Autonomy," launched by the United States in 2023. This declaration sets "voluntary guidelines describing best practices for use of AI in a military context" and aims to "put in place measures to increase transparency, communication, and reduce risks of inadvertent conflict and escalation."

The declaration represents an unprecedented level of international cooperation on military AI governance, extending well beyond traditional alliance structures to include partners across multiple regions. The broad endorsement demonstrates growing recognition that AI military applications require international coordination to prevent misunderstandings and reduce escalation risks.

Joint Training and Operational Integration

NATO's Unified Vision exercises demonstrate practical approaches to AI integration across alliance operations. The 2023 iteration involved "18 Allies, Sweden, numerous NATO entities and industry experts" testing abilities to "collect, process, exploit and share intelligence from seabed to space." The exercise specifically focused on incorporating "Emerging and Disruptive Technologies – such as artificial intelligence and automation tools – in its intelligence process."

These exercises reveal both the promise and challenges of AI integration. While AI can enhance information processing and analysis capabilities, successful integration requires extensive preparation and standardization to ensure systems from different countries can effectively collaborate.

Future Alliance Models: Adapting to AI-Native Operations

Specialization and Division of Labor

The AI age is driving NATO toward a more specialized approach to capability development where different allies lead in different technological domains rather than duplicating capabilities. This model recognizes that few allies can match the comprehensive AI development resources of major powers, but many can contribute specialized expertise in specific areas.

Estonia's focus on cyber defense and digital resilience, Singapore's smart nation initiatives, and Ukraine's wartime AI innovation demonstrate how different approaches to AI specialization can enhance collective capabilities while playing to national strengths.

Real-Time Coordination Systems

Future alliance operations may require AI systems capable of real-time coordination across multiple national militaries without human intervention in time-critical situations. This represents a fundamental shift from current procedures where human operators maintain control over cross-border military coordination.

NATO's planning for such capabilities focuses on ensuring "autonomous systems programmed to operate together" while maintaining human oversight for critical decisions. This approach attempts to capture AI speed advantages while preserving human accountability for alliance decisions.

Public-Private Integration

The increasing role of private companies in developing AI capabilities forces NATO to adapt traditional alliance structures that were designed for government-to-government cooperation.

DIANA's approach of engaging directly with start-ups and innovative companies represents one model for integrating private sector innovation into alliance frameworks.

The NATO Innovation Fund's investment approach recognizes that critical AI technologies may be developed outside traditional defense contractors, requiring new mechanisms for incorporating commercial innovation into alliance capabilities while maintaining security and control.

Persistent Challenges: The Limits of Cooperation

Attribution and Accountability

One of the most difficult challenges facing AI-enabled alliance operations involves maintaining accountability when autonomous systems from multiple countries operate together. If an AI system from one ally makes a targeting recommendation that another ally's system acts upon, determining responsibility for outcomes becomes extremely complex.

NATO's emphasis on "explainability and traceability" in its responsible AI principles attempts to address this challenge, but technical implementation remains difficult when systems from different countries may use different algorithms and training data.

Classification and Operational Security

The fundamental challenge of "classification by compilation" continues to constrain AI cooperation within the alliance. Traditional security frameworks struggle to adapt to AI systems that can synthesize insights from multiple unclassified sources to reveal classified information.

This problem becomes particularly acute in intelligence applications where AI systems might reveal sources and methods through their analysis patterns, even when the underlying data appears unclassified. NATO is working to develop new approaches to information security that can accommodate AI systems' analytical capabilities while protecting sensitive information.

Technological Dependencies

The concentration of advanced AI capabilities in a few countries creates potential vulnerabilities for alliance cohesion. If key allies restrict access to critical AI technologies, other alliance members may be forced to develop alternative capabilities or accept technological dependence that could constrain their sovereign decision-making.

Recent U.S. export controls on AI technologies demonstrate this challenge, as even close allies may find themselves subject to restrictions that limit their ability to develop independent AI capabilities.

Key Insights: The Transformation of Alliance Relationships

The integration of AI into alliance operations is fundamentally transforming the nature of international cooperation from shared values and common threats to shared algorithms and compatible systems. Success requires not just political agreement on objectives but technical standardization of decision-making processes themselves.

This transformation creates new forms of interdependence where allies must trust not just each other's intentions but each other's algorithms. The ability to maintain alliance cohesion while managing this technological interdependence will largely determine the effectiveness of collective defense in the AI age.

The experience suggests that successful AI integration requires:

1. **Technical standardization** that goes far beyond traditional interoperability to encompass decision-making algorithms themselves
2. **New burden-sharing models** that account for technological contributions rather than just traditional military spending
3. **Flexible security frameworks** that can accommodate AI systems' analytical capabilities while protecting sensitive information
4. **Continuous adaptation mechanisms** that can keep pace with rapid technological development

The stakes are particularly high because AI offers potential advantages to authoritarian systems that can implement centralized standards more quickly than democratic alliance. NATO's success in developing effective AI cooperation frameworks may determine whether democratic values can be preserved in an AI-dominated strategic environment.

As one NATO official noted, "AI is too important to be relegated to discussions on NATO's operational and tactical levels. NATO's political and strategic leaders need to lay out a vision for AI's role in the Alliance, before it is too late." The alliance's ability to meet this challenge will largely determine its relevance in the 21st century.

Sources and References

1. NATO Official Website: https://www.nato.int/cps/en/natohq/news_227234.htm - "NATO releases revised AI strategy" (July 10, 2024)
2. NATO DIANA Website: https://www.diana.nato.int/ - Defence Innovation Accelerator for the North Atlantic
3. NATO Official Text: https://www.nato.int/cps/en/natohq/official_texts_227237.htm - "Summary of NATO's revised Artificial Intelligence (AI) strategy" (July 10, 2024)
4. NATO News: https://www.nato.int/cps/en/natohq/news_223648.htm - "DIANA doubles the size of its transatlantic network" (March 14, 2024)
5. NATO Official Text: https://www.nato.int/cps/en/natohq/official_texts_208374.htm - "NATO's Data and Artificial Intelligence Review Board" (October 13, 2022)

6. Center for Strategic and International Studies: https://www.csis.org/analysis/calibrating-natos-vision-ai-enabled-decision-support (October 11, 2024)
7. NATO Review: https://www.nato.int/docu/review/articles/2021/10/25/an-artificial-intelligence-strategy-for-nato/index.html (October 25, 2021)
8. Council on Foreign Relations: https://www.cfr.org/blog/what-know-about-new-us-ai-diffusion-policy-and-export-controls (January 13, 2025)
9. Center for Strategic and International Studies: https://www.csis.org/analysis/understanding-us-allies-current-legal-authority-implement-ai-and-semiconductor-export (April 2, 2025)
10. DefenseScoop: https://defensescoop.com/2024/10/29/nearly-60-militaries-endorse-us-pledge-responsible-ai-cdao/ (October 29, 2024)
11. Texas National Security Review: https://tnsr.org/2020/03/allies-and-artificial-intelligence-obstacles-to-operations-and-decision-making/ (April 16, 2025)
12. Breaking Defense: https://breakingdefense.com/2024/07/nato-updates-ai-strategy-for-the-age-of-chatgpt/ (July 25, 2024)
13. NATO News: https://www.nato.int/cps/en/natohq/news_211498.htm - "NATO starts work on Artificial Intelligence certification standard" (February 7, 2023)
14. The Washington Post: https://www.washingtonpost.com/technology/2025/01/13/ai-export-controls-chips-china/ (January 14, 2025)
15. NATO Parliamentary Assembly: https://www.nato-pa.int/document/2024-nato-and-ai-report-clement-058-stc (2024)

Chapter 13: Emerging Strategic Cultures: The Next Generation

Projecting the Evolution Through 2040

Opening Vision: The 2035 Battlefield

Lieutenant Colonel Sarah Chen commands a multi-domain task force operating along the Taiwan Strait in 2035. Her human-AI command team processes thousands of data streams simultaneously—satellite imagery, quantum-encrypted communications, social media sentiment analysis, and biometric feedback from her soldiers' neural interfaces. Within seconds, AI algorithms assess terrain, readiness, doctrine, and historical precedents to generate tactical options. But the final decision remains hers: a product of strategic intuition honed through years of human-AI collaborative training.

This scenario, emerging from current technological trajectories and documented military developments, represents more than tactical evolution—it signals the birth of post-human strategic cultures where artificial and human intelligence merge seamlessly in strategic decision-making.

Generational Strategic Shifts

Post-Human Strategic Cultures

We stand at the threshold of a fundamental transformation in how nations conceptualize and conduct strategic affairs. The current generation of military leaders, born into analog strategic cultures, will soon be replaced by officers who have never known warfare without AI assistance. This transition represents more than technological adoption—it constitutes a civilizational shift toward post-human strategic cultures.

The U.S. Army's creation of new artificial intelligence-focused occupational specialties in 2025, including the 49B enlisted specialty and parallel warrant officer tracks, signals institutional recognition of this transformation. As these AI-native officers ascend to senior leadership positions over the next two decades, they will bring fundamentally different assumptions about the role of human judgment versus machine analysis in strategic decision-making.

Unlike current leaders who view AI as a tool to be carefully integrated, the emerging generation will consider human-AI collaboration as the natural state of strategic affairs. Their strategic cultures will be built around what DARPA researchers term "human-machine teaming" (HMT), where decision-making emerges from seamless collaboration between human creativity and machine precision.

Current military education is already adapting to this reality. Professional Military Education (PME) institutions are integrating AI tools directly into their curricula, with faculty using large language models to generate customized strategic scenarios and AI-enabled wargaming systems that simulate expert decision-making. As one military educator noted in 2025, "The institutions that thrive will be those that teach students to leverage AI effectively—using it to augment rather than replace human judgment."

This educational transformation is producing officers who think differently about strategic problems. Where previous generations might have relied on historical analogies and institutional memory, emerging leaders will instinctively turn to AI-assisted pattern recognition across vast datasets of strategic precedents. They will expect real-time decision support systems that can process geopolitical, economic, and military variables simultaneously.

The implications extend beyond individual decision-making to organizational culture. Post-human strategic cultures will feature flattened command hierarchies where AI systems enable junior officers to access strategic-level information and analysis previously reserved for senior leadership. This democratization of strategic intelligence will require new forms of institutional control and coordination.

Cultural DNA Changes

Strategic cultures encode nations' deepest assumptions about the nature of conflict, the role of military force, and the relationship between war and politics. As AI becomes embedded in these cultures, fundamental changes are occurring in how nations conceptualize strategic competition.

The traditional American preference for technological solutions to strategic problems is evolving into an AI-centric worldview where algorithmic superiority becomes synonymous with strategic advantage. The Pentagon's $1.8 billion AI budget for fiscal 2025, representing consistent growth from $874 million in fiscal 2022, reflects this cultural shift toward viewing AI capabilities as the primary determinant of military effectiveness.

China's "intelligentization" doctrine represents perhaps the most comprehensive reimagining of strategic culture for the AI age. The People's Liberation Army's goal to "accelerate the integrated development of mechanization, informatization, and intelligentization" by 2027 embeds AI directly into the nation's strategic DNA. Chinese strategists envision warfare where "human cognition can no longer keep pace with speed of decision-making," requiring strategic cultures built around what they term "battlefield singularity."

This represents a departure from traditional Chinese strategic thought, which emphasized human factors and patient long-term planning. The new intelligentized strategic culture accepts AI acceleration of strategic timelines and embraces machine-speed decision cycles that previous Chinese doctrine would have considered reckless.

European strategic cultures are evolving differently, emphasizing what the European Union terms "digital sovereignty" and responsible AI frameworks. NATO's revised AI strategy from July 2024 establishes six principles of responsible use—lawfulness, responsibility, explainability, reliability, governability, and bias mitigation—that embed ethical constraints directly into alliance strategic culture.

Institutional Evolution

Military institutions worldwide are restructuring themselves for AI integration. The U.S. military's transition from the Joint Artificial Intelligence Center to the Chief Digital and Artificial Intelligence Office (CDAO) represents more than organizational reshuffling—it signals institutional recognition that AI requires permanent, enterprise-wide integration rather than project-based adoption.

Military academies and war colleges are rebuilding their curricula around human-AI collaboration. The Naval War College now employs CustomGPTs that simulate expertise in U.S.-China strategic competition and historical great-power rivalry, enabling students to engage in sophisticated strategic wargaming against AI opponents that model adversary decision-making patterns.

These educational changes are producing officers with fundamentally different cognitive approaches to strategic problems. Traditional military education emphasized pattern recognition through historical study and case analysis. AI-augmented education enables officers to process vastly larger datasets of strategic precedents while engaging with AI systems that can simulate alternative strategic perspectives in real-time.

The institutional challenge lies in maintaining strategic coherence while enabling AI-driven adaptation. Military organizations must develop what researchers term "AI certification standards" to ensure human-AI teams can operate effectively across alliance structures. NATO's Defense Innovation Accelerator (DIANA) and €1 billion Innovation Fund represent institutional attempts to coordinate this transformation across multiple national strategic cultures.

Technology Convergence Trends

Quantum-AI Integration

The convergence of quantum computing and artificial intelligence represents the most transformative technological development since nuclear weapons. China's massive investments in quantum research, including billions allocated through consecutive five-year plans since 2006, position it to achieve quantum advantages that could revolutionize strategic calculation.

Chinese researchers claimed in 2021 that their quantum computer completed tasks in one hour that would require eight years on conventional supercomputers. While these claims remain difficult to verify independently, the strategic implications are profound. Quantum-enhanced AI systems could enable computational capabilities that render current strategic assumptions obsolete.

The military applications of quantum-AI convergence span multiple domains. Quantum sensors could detect previously undetectable stealth aircraft and submarines, fundamentally altering force projection capabilities. Quantum communication networks could provide unhackable command

and control systems. Most significantly, quantum computers could break current encryption standards while enabling new forms of algorithmic warfare that operate at speeds beyond human comprehension.

The U.S. Defense Intelligence Agency's 2025 threat assessment warns that quantum technologies are "nearing battlefield relevance" with China and Russia expanding "city-scale quantum networks" and developing detection tools that could undermine traditional stealth advantages. This assessment indicates that quantum-AI convergence will reshape strategic balance within the current decade rather than remaining a distant possibility.

Brain-Computer Interfaces

Perhaps no technology represents the literal convergence of human and artificial intelligence more clearly than brain-computer interfaces (BCIs). DARPA's Next-Generation Nonsurgical Neurotechnology (N^3) program, with $104 million in funding, aims to develop systems that enable direct neural control of military systems without surgical implantation.

The strategic implications extend far beyond individual augmentation. BCIs could enable commanders to interface directly with AI strategic planning systems, accessing machine-generated options through thought alone while contributing human intuition and creativity to algorithmic analysis. Early research suggests such systems could enable control of drone swarms through neural commands, potentially allowing single operators to coordinate hundreds of autonomous systems.

China's investment in what it terms "polymeric technology combining brain science, biological technology, and AI" suggests a different approach to human-machine integration. Chinese research emphasizes collective neural interfaces that could enable new forms of command and control where multiple human operators share cognitive load with AI systems.

The strategic culture implications are profound. Nations that successfully integrate BCIs into military operations could achieve decision-making speeds that render traditional command structures obsolete. The ethical and legal frameworks governing such integration remain underdeveloped, creating strategic advantages for nations willing to accept greater risks in human enhancement.

Autonomous Ecosystems

The evolution toward fully autonomous military ecosystems represents the culmination of AI integration in strategic affairs. Current developments in drone swarming, where AI systems coordinate hundreds of autonomous platforms simultaneously, provide a glimpse of future warfare where human oversight becomes optional rather than essential.

The 2024 testing of countermeasures against swarms of up to 40 drones by the U.S. Army demonstrates both the advancing capability of autonomous systems and the challenge of defending against them. Chinese advances in autonomous vehicle deployment, including robot dogs armed with machine guns delivered by drones, suggest rapid progression toward autonomous ecosystems that can operate across multiple domains without human intervention.

These autonomous ecosystems will enable entirely new forms of strategic competition. Nations could deploy autonomous systems for persistent surveillance, area denial, and graduated escalation without risking human casualties or diplomatic confrontation. The strategic challenge lies in maintaining human agency over strategic decisions while leveraging the speed and scale advantages of autonomous systems.

Future Operational Concepts

Hyperwar Scenarios

Military strategists have begun conceptualizing "hyperwar"—conflicts conducted at machine speed where human decision-making becomes a limiting factor rather than an advantage. Former Marine Corps General John R. Allen and AI entrepreneur Amir Husain describe warfare fought by "swarms of unmanned, autonomous weapons across land, air, sea, space and cyber" using "artificial intelligence-based algorithms to make decisions" in "real-time and independent of human decision making."

The implications for strategic culture are profound. Hyperwar scenarios compress strategic decision-making from days or weeks to minutes or seconds. Traditional concepts of escalation control, based on human deliberation and diplomatic communication, may become obsolete when AI systems can execute strategic actions faster than humans can recognize and respond to them.

Current military AI spending, which reached $9.2 billion in 2023 with forecasts approaching $61.08 billion by 2034, reflects institutional preparation for hyperwar scenarios. The Pentagon's Replicator program, allocated $500 million in fiscal 2024, aims to field "multiple thousands of all-domain autonomous systems" by August 2025, indicating that hyperwar capabilities are transitioning from conceptual to operational.

The strategic culture challenge lies in maintaining human control over strategic outcomes while operating at machine speed. Nations must develop what researchers term "meaningful human control" frameworks that preserve human agency over strategic decisions while enabling AI systems to execute tactical actions at hyperwar speeds.

Cognitive Dominance

Chinese strategic thinking increasingly emphasizes what PLA theorists call the "cognitive space where effective decision-making provides decisive advantage." This concept envisions warfare where victory depends on controlling adversary perception and decision-making rather than destroying physical capabilities.

AI-enhanced cognitive warfare capabilities enable unprecedented manipulation of information environments. Chinese deployment of "sophisticated machine learning models to create personalized propaganda aimed at Taiwan" demonstrates operational applications of cognitive dominance strategies. These systems can generate targeted disinformation campaigns adapted to individual psychological profiles derived from social media analysis.

The strategic implications extend beyond information warfare to fundamental questions about the nature of strategic competition. When AI systems can generate convincing false intelligence, manipulate social media sentiment, and create deepfake evidence of adversary actions, traditional concepts of strategic assessment and decision-making become problematic.

Future strategic cultures must develop new frameworks for operating in environments where information authenticity cannot be assumed. This requires strategic leaders capable of making critical decisions based on incomplete and potentially manipulated information while leveraging AI systems to detect and counter adversary cognitive warfare campaigns.

Multi-Domain Precision Warfare

Chinese strategists envision "multi-domain precision warfare" that integrates AI capabilities across cyber, space, and kinetic domains simultaneously. This concept represents evolution beyond traditional sequential operations toward synchronized effects across all domains of conflict.

The technological foundation exists for such operations. AI systems can coordinate cyber attacks, satellite operations, and kinetic strikes in real-time while adapting to adversary responses faster than human commanders can process information. Chinese research into "system destruction warfare" targeting "vulnerable links between sensors and platforms" suggests strategic concepts that treat adversary military systems as integrated networks vulnerable to coordinated multi-domain disruption.

The strategic culture implications require new approaches to alliance coordination and escalation management. When AI systems can execute multi-domain operations simultaneously, traditional concepts of graduated response and alliance consultation may become inadequate for managing strategic competition.

Cultural Evolution Patterns

Speed vs. Reflection Balance

The fundamental tension in emerging strategic cultures lies between AI-enabled speed and human reflection. Military organizations must balance the competitive advantages of machine-speed decision-making against the strategic wisdom that emerges from deliberate human analysis.

The U.S. Army's concept of mission command is evolving into what researchers describe as "dialogue between man and machine" where human commanders provide strategic intent while AI systems execute tactical details. This approach attempts to preserve human strategic agency while leveraging machine capabilities for operational execution.

However, the speed of modern conflicts may render such distinctions artificial. When adversary AI systems can complete strategic assessment and response cycles in minutes, the luxury of human deliberation may become a competitive disadvantage rather than a strategic asset.

Successful strategic cultures will develop institutional mechanisms for maintaining reflection capabilities while operating at AI speeds. This might include AI systems designed to preserve space for human deliberation, automated escalation controls that require human authorization, and training programs that enhance human strategic judgment rather than replacing it with algorithmic analysis.

Democratic vs. Authoritarian AI Cultures

The integration of AI into strategic affairs reveals fundamental differences between democratic and authoritarian approaches to human-machine collaboration. Authoritarian systems potentially possess advantages in AI integration due to centralized decision-making and reduced oversight requirements.

China's strategic culture accepts "risks, costs and long-term commitments for low-probability, high-return innovation" in AI development, enabling rapid experimentation with autonomous systems and human-AI integration. The PLA's development of AI-enabled autonomous weapons proceeds without the democratic oversight requirements that constrain U.S. and allied programs.

Conversely, democratic strategic cultures must maintain public accountability and alliance consultation while competing against adversaries that face fewer constraints on AI experimentation. The challenge lies in preserving democratic values while achieving competitive AI capabilities.

The solution may lie in leveraging democratic advantages in innovation and adaptation. Open societies excel at generating technological innovation and adapting to changing circumstances—capabilities that may prove decisive in the long-term development of strategic AI systems.

Alliance vs. National AI Cultures

The development of AI strategic capabilities creates new tensions between national advantages and alliance coordination. Nations investing heavily in AI capabilities may be reluctant to share algorithms and data with allies, potentially weakening collective security while strengthening individual capabilities.

NATO's efforts to develop "Alliance-wide AI Testing, Evaluation, Verification & Validation" frameworks attempt to balance these competing demands. The alliance seeks to enable interoperability while preserving national competitive advantages in AI development.

The strategic culture challenge lies in developing shared approaches to human-AI integration across multiple national traditions and legal systems. Different nations bring distinct perspectives on human agency, technological risk, and strategic decision-making that must be reconciled in alliance AI systems.

Successful alliance strategic cultures will likely feature specialization rather than replication, with different nations leading in different AI domains while contributing to collective capabilities. This approach maximizes innovation while preserving alliance coordination.

Potential Discontinuities

Technological Breakthroughs

The nonlinear nature of AI development creates potential for sudden technological breakthroughs that could rapidly alter strategic balance. Quantum computing advances, artificial general intelligence development, or breakthrough applications of brain-computer interfaces could provide decisive advantages to early adopters.

The strategic challenge lies in preparing for discontinuous change while investing in current capabilities. Nations must balance research into potentially revolutionary technologies against requirements for immediate military effectiveness.

China's willingness to accept "low-probability, high-return innovation" may provide advantages in pursuing breakthrough technologies, while democratic systems may excel at incremental innovation and adaptation. The strategic outcome will depend on whether revolutionary or evolutionary approaches prove more effective in AI development.

Generational Replacement

The most profound discontinuity may emerge from generational replacement of strategic leadership. As officers who trained with AI from the beginning of their careers ascend to senior positions, they will bring fundamentally different assumptions about human-AI collaboration.

This generational transition could occur rapidly, particularly if conflict experience accelerates promotion timelines. Young officers who demonstrated exceptional capability in human-AI collaboration during crisis situations could advance quickly to strategic leadership positions.

The strategic culture implications could be dramatic. AI-native leaders might accept levels of machine autonomy that current leadership considers unacceptable, potentially creating rapid shifts in strategic doctrine and force structure.

Crisis-Driven Evolution

Major conflicts could accelerate strategic culture transformation by demonstrating the effectiveness or limitations of AI-enabled operations. The ongoing conflict in Ukraine has already shown how rapidly military organizations can adapt AI technologies when survival depends on innovation.

Future conflicts involving major powers could produce even more dramatic adaptations. Strategic cultures that prove capable of rapid AI integration under combat conditions may achieve decisive advantages over those constrained by peacetime institutional processes.

The risk lies in crisis-driven evolution outpacing thoughtful integration. Strategic cultures might adopt AI capabilities without developing adequate frameworks for human oversight, potentially creating systemic risks that extend beyond immediate military effectiveness.

Regional Variations

East Asian AI Military Cultures

East Asian nations are developing distinct approaches to AI strategic integration that reflect regional strategic cultures and technological capabilities. China's comprehensive "intelligentization" program represents the most ambitious attempt to rebuild strategic culture around AI capabilities.

Japan's approach emphasizes defensive AI applications and alliance integration with the United States, reflecting constitutional constraints on offensive capabilities. South Korea's focus on AI-enhanced early warning systems reflects the compressed strategic geography of the Korean Peninsula.

These regional variations suggest that AI strategic cultures will develop along existing cultural and institutional lines rather than converging toward universal approaches. Regional strategic traditions will shape how nations integrate AI capabilities and define human-machine relationships.

European Regulatory Leadership

European strategic cultures are attempting to shape global AI development through regulatory frameworks rather than competing primarily on technological capabilities. The European Union's AI Act and emphasis on "digital sovereignty" represent efforts to define responsible AI development standards that could influence global norms.

This approach leverages European strengths in international law and regulatory coordination while acknowledging limitations in competing directly with U.S. and Chinese AI capabilities. European strategic cultures may achieve influence by defining ethical frameworks that other nations adopt for legitimacy purposes.

The effectiveness of this approach will depend on whether regulatory standards can shape technological development or whether technological capabilities ultimately determine regulatory outcomes.

Middle Power Specialization

Middle powers are developing AI strategic cultures based on specialization rather than comprehensive capabilities. Canada's 2024 Defence AI Strategy exemplifies this approach by focusing on specific AI domains where Canadian capabilities can contribute to broader alliance structures.

This specialization strategy acknowledges resource constraints while seeking to maintain strategic relevance through niche capabilities. Middle powers may achieve disproportionate influence by leading in specific AI applications that larger powers require for comprehensive capabilities.

The strategic culture challenge lies in maintaining coherent strategic vision while pursuing specialized capabilities. Middle powers must balance national sovereignty with alliance integration in ways that preserve strategic autonomy while contributing to collective security.

Measurement and Assessment

Success Metrics for AI Strategic Transformation

Measuring success in AI strategic culture transformation requires new frameworks that account for both technological capabilities and organizational adaptation. Traditional metrics of military effectiveness may prove inadequate for assessing human-AI collaborative systems.

Potential metrics include the speed and accuracy of strategic decision-making, the ability to operate effectively in contested information environments, and the capacity to adapt to adversary AI capabilities. Organizational metrics might assess the integration of AI-native personnel, the effectiveness of human-AI training programs, and the development of appropriate oversight mechanisms.

The challenge lies in developing metrics that capture qualitative changes in strategic culture rather than simply measuring technological adoption. Nations may possess advanced AI capabilities while failing to develop strategic cultures capable of employing them effectively.

Warning Indicators of Problematic Development

Strategic leaders must develop early warning systems for problematic AI integration that could undermine strategic effectiveness or human control. Warning indicators might include over-reliance on AI analysis without human oversight, failure to maintain human strategic skills, or loss of alliance interoperability due to incompatible AI systems.

Organizational indicators could include degradation of human strategic thinking capabilities, excessive confidence in AI predictions, or failure to maintain contingency plans for AI system failures. Cultural indicators might involve loss of strategic patience as leaders become accustomed to machine-speed analysis.

The development of these warning systems requires honest assessment of AI limitations and potential failure modes. Strategic cultures that maintain awareness of AI weaknesses while leveraging AI strengths will prove more resilient than those that develop excessive dependence on machine capabilities.

Benchmarks for Human-AI Integration Effectiveness

Effective human-AI integration requires benchmarks that assess the quality of collaboration rather than simply the presence of AI systems. These benchmarks might include the ability of human commanders to override AI recommendations when appropriate, the capacity of AI systems to explain their analysis to human users, and the effectiveness of human-AI teams in novel strategic situations.

Training benchmarks could assess the ability of strategic leaders to work effectively with AI systems while maintaining independent judgment. Operational benchmarks might measure the speed and accuracy of human-AI collaborative decision-making compared to either human-only or AI-only approaches.

The development of these benchmarks requires extensive experimentation with human-AI teams in realistic strategic scenarios. Military education institutions and strategic planning organizations must create environments for testing and refining human-AI collaboration approaches.

Key Insight Development

The Post-Human Strategic Revolution

We are witnessing the emergence of post-human strategic cultures where the distinction between human and artificial intelligence becomes increasingly meaningless for strategic purposes. This transformation represents more than technological adoption—it constitutes a fundamental shift in how humans conceptualize and conduct strategic competition.

The implications extend far beyond military affairs to questions of human agency, democratic governance, and the nature of strategic wisdom. As AI systems become capable of strategic analysis and decision-making that rivals or exceeds human capabilities, traditional concepts of strategic leadership and institutional accountability require fundamental reconceptualization.

Nations that successfully navigate this transformation will develop strategic cultures that seamlessly integrate human wisdom with machine intelligence while preserving human agency over ultimate strategic decisions. Those that fail to adapt risk strategic irrelevance in a post-human strategic environment.

The Imperative for Conscious Design

Unlike previous technological transformations that emerged gradually over decades, the AI strategic revolution is occurring at unprecedented speed. The window for conscious strategic culture design is rapidly closing as technological capabilities outpace institutional adaptation.

Strategic leaders have perhaps five years to shape how their nations integrate AI into strategic affairs before approaches become locked in by institutional inertia, technological constraints, and competitive pressures. After this window closes, strategic cultures will be shaped by technological momentum rather than deliberate choice.

The stakes could not be higher. The choices made in the next few years will determine whether AI enhances human strategic wisdom or replaces it, whether democratic societies can compete effectively against authoritarian AI development, and whether strategic stability can be maintained in an AI-accelerated international system.

Success requires more than technological investment—it demands conscious attention to preserving human agency, maintaining democratic accountability, and ensuring that AI capabilities serve human flourishing rather than replacing human judgment in matters of war and peace.

Verifiable References and Sources

1. **Army Creates New AI Military Occupational Specialty** - Military.com, July 2, 2025
    - https://www.military.com/daily-news/2025/07/02/army-creating-new-artificial-

intelligence-focused-occupational-specialty-and-officer-field.html

2. **Pentagon AI Budget Request FY2025** - DefenseScoop, March 11, 2024
   - https://defensescoop.com/2024/03/11/pentagon-ai-budget-request-2025/
3. **Military AI Spending Growth** - TIME, March 29, 2024
   - https://time.com/6961317/ai-artificial-intelligence-us-military-spending/
4. **Global AI Military Market Analysis** - Research and Markets, May 9, 2025
   - https://www.globenewswire.com/news-release/2025/05/09/3078053/0/en/61-08-Bn-Artificial-Intelligence-in-Military-Market-Opportunities-and-Strategies-to-2034-Fueled-by-AI-Adoption-and-Defense-Budget-Increases-Amid-Privacy-and-Talent-Concerns.html
5. **Pentagon Replicator Program Details** - Washington Times, May 8, 2024
   - https://www.washingtontimes.com/news/2024/may/8/pentagon-reveals-new-details-on-how-its-spending-f/
6. **China's Military AI and Intelligentization Strategy** - Center for New American Security, February 1, 2024
   - https://www.cnas.org/publications/congressional-testimony/military-artificial-intelligence-the-peoples-liberation-army-and-u-s-china-strategic-competition
7. **PLA AI Development Analysis** - ICS Research Blog, January 13, 2021
   - https://icsin.org/blogs/2021/01/07/a-critical-outlook-on-plas-ai-development-philosophy/
8. **China's AI Military Applications** - HPC Wire, January 8, 2025
   - https://www.hpcwire.com/2025/01/07/dods-2024-china-report-highlights-plans-for-ai-and-quantum-in-military-use/
9. **China PLA Generative AI Usage** - Recorded Future, December 2024
   - https://www.recordedfuture.com/research/artificial-eyes-generative-ai-chinas-military-intelligence
10. **DARPA Brain-Computer Interface Research** - DARPA N^3 Program
    - https://www.darpa.mil/research/programs/next-generation-nonsurgical-neurotechnology
11. **BCI Military Applications Assessment** - RAND Corporation, 2020
    - https://www.rand.org/pubs/research_reports/RR2996.html
12. **US Military Mind-Reading Technology** - MIT Technology Review, August 22, 2024
    - https://www.technologyreview.com/2019/10/16/132269/us-military-super-soldiers-control-drones-brain-computer-interfaces/
13. **Quantum Computing Military Applications** - The Quantum Insider, May 27, 2025
    - https://thequantuminsider.com/2025/05/27/u-s-defense-intelligence-flags-

rivals-growing-military-use-of-quantum-tech/

14. **US-China Quantum Arms Race** - MIT Technology Review, August 22, 2024
    - https://www.technologyreview.com/2019/01/03/137969/us-china-quantum-arms-race/
15. **China Quantum Computing Leadership** - AFCEA International
    - https://www.afcea.org/signal-media/us-trails-china-decade-quantum-computing
16. **Hyperwar Concept Analysis** - C4ISRNET, August 7, 2017
    - https://www.c4isrnet.com/dod/2017/08/07/emerging-hyperwar-signals-ai-fueled-machine-waged-future-of-conflict/
17. **Future Warfare Autonomous Technologies** - Small Wars Journal, January 28, 2025
    - https://smallwarsjournal.com/2025/01/28/the-future-of-warfare-autonomous-technologies-in-modern-conflict/
18. **AI Hyperwar Scenarios** - Big Think
    - https://bigthink.com/the-future/hyperwar-ai-military-warfare/
19. **Military Drone Swarm Technology** - Total Military Insight, June 28, 2024
    - https://totalmilitaryinsight.com/military-drone-swarm-technology/
20. **Human-Machine Teaming Research** - Atlantic Council, August 17, 2023
    - https://www.atlanticcouncil.org/in-depth-research-reports/report/how-modern-militaries-are-leveraging-ai/
21. **Army AI Research Programs** - Next Gov, April 27, 2020
    - https://www.nextgov.com/emerging-tech/2020/04/army-wants-study-how-humans-team-aiand-vice-versa/164770/
22. **DARPA Human-AI Teams Research** - Military Aerospace
    - https://www.militaryaerospace.com/computers/article/55126930/artificial-intelligence-ai-machine-learning-military-operations
23. **Trust in Human-Machine Teaming** - War on the Rocks, March 7, 2023
    - https://warontherocks.com/2023/03/miscalibration-of-trust-in-human-machine-teaming/
24. **AI in Military Education** - Military.com, April 22, 2024
    - https://www.military.com/daily-news/2024/04/22/use-of-ai-war-games-could-change-military-strategy.html
25. **PME AI Integration** - Small Wars Journal, May 7, 2025
    - https://smallwarsjournal.com/2025/05/07/embracing-the-inevitable/
26. **AI in Military Classroom** - War on the Rocks, March 31, 2025
    - https://warontherocks.com/2025/04/peering-into-the-future-of-artificial-intelligence-in-the-military-classroom/

27. **Naval Strategic Culture Evolution** - U.S. Naval Institute Proceedings, November 2022
    - https://www.usni.org/magazines/proceedings/2022/november/how-rebalance-navys-strategic-culture
28. **Leadership in AI Warfare** - Federal News Network, April 3, 2025
    - https://federalnewsnetwork.com/commentary/2025/04/leadership-trust-and-innovation-the-keys-to-ais-role-in-modern-warfare-2/
29. **Army Cyber AI Research** - AFCEA International
    - https://www.afcea.org/signal-media/cyber-edge/army-cyber-researchers-develop-ai-red-teaming-capabilities
30. **AI Arms Race Overview** - Wikipedia, July 2025
    - https://en.wikipedia.org/wiki/Artificial_intelligence_arms_race
31. **Generative World Order Analysis** - Goldman Sachs, December 14, 2023
    - https://www.goldmansachs.com/insights/articles/the-generative-world-order-ai-geopolitics-and-power
32. **Military AI Information Operations** - Modern War Institute, October 25, 2024
    - https://mwi.westpoint.edu/persuade-change-and-influence-with-ai-leveraging-artificial-intelligence-in-the-information-environment/

Chapter 14: Risks, Perils, and Unintended Consequences

When AI Strategic Systems Fail or Conflict

Opening Warning: The Reality of AI Military Failures

In September 1983, Lieutenant Colonel Stanislav Petrov made a decision that likely saved the world. When Soviet early warning systems detected what appeared to be five incoming American intercontinental ballistic missiles, Petrov chose not to report the alert up the chain of command. His reasoning was simple: if the United States were to launch a nuclear attack, it would likely involve hundreds of missiles, not just five. The computer system had malfunctioned, mistaking sunlight reflecting off clouds for missile launches. Petrov's human judgment overruled the machine's "certainty," preventing what could have been nuclear war.

Today, as artificial intelligence systems become increasingly integrated into military decision-making, the margin for such human intervention is rapidly shrinking. The Department of Defense's 2024 Framework for Advancing AI Governance acknowledges that "decision cycles are measured in seconds, not days," while maintaining that humans will remain "in the loop" for critical decisions. Yet experts warn that the very speed that makes AI attractive in military applications may also be its greatest danger.

The stakes of getting this wrong are existential. As Paul Scharre, a leading expert on autonomous weapons, warned in 2025: "What is the training data set we have for nuclear war?" The question cuts to the heart of a fundamental paradox: AI systems are trained on historical data to make predictions about unprecedented situations, potentially leading to catastrophic miscalculations in scenarios humanity has never experienced.

Systemic Risks Analysis

AI Strategic Mishaps: The Scale of Current Deployment

The proliferation of AI in military systems has accelerated dramatically. The Pentagon identified over 180 instances of AI military applications as of 2024, ranging from intelligence analysis to battlefield assessment. The U.S. military AI budget increased from $4.6 billion in 2022 to $9.2 billion in 2023, with forecasts reaching $38.8 billion by 2028. Yet this rapid expansion has occurred largely without comprehensive testing of how these systems might interact or fail in high-stress environments.

Military officials acknowledge significant classification challenges where "AI constantly creating new analyses" outpaces security review processes. This creates a dangerous gap where AI systems may be making recommendations or taking actions based on information that hasn't been properly vetted by human oversight. The International Committee of the Red Cross warned in September 2024 that AI-based decision support systems for target selection carry "grave risk" of actually exacerbating civilian suffering rather than reducing it.

Flash War Scenarios: When Speed Becomes the Enemy

The Center for Security and Emerging Technology concluded in December 2023 that AI acceleration creates "crisis scenarios where tensions are high and communications are strained," potentially leading to "authorized nuclear exchange" within hours rather than days. This compression of decision-making timelines represents a fundamental shift in the nature of strategic stability.

Consider the implications: during the Cuban Missile Crisis, President Kennedy had thirteen days to deliberate with advisors, weigh options, and pursue diplomatic solutions. In an AI-accelerated conflict, leaders might have thirteen minutes—or less. The RAND Corporation's 2018 analysis concluded that AI could "upend concepts of nuclear deterrence that have helped spare the world from nuclear war since 1945."

Cross-Domain Escalation Dynamics

AI systems operate across multiple domains simultaneously—cyber, space, information, and kinetic—creating unprecedented possibilities for cascading failures. The Stockholm International Peace Research Institute warned in 2025 that "non-nuclear applications of military AI may compress decision-making timelines, potentially increasing miscalculation risks during a crisis."

The danger lies not just in direct AI control of weapons, but in how AI systems influence human decision-makers. As research published in 2024 showed, AI potentially undermines "fundamental pillars of international security such as deterrence" by making "rationality-based deterrence logic increasingly untenable." When machines process information faster than humans can comprehend, the foundations of strategic stability—which depend on rational actors making calculated decisions—begin to erode.

Technical Vulnerabilities

Algorithmic Bias in Life-and-Death Decisions

Perhaps no risk is more insidious than algorithmic bias in military AI systems. Research published by the International Committee of the Red Cross in September 2024 revealed that AI picking up "inadvertent biases from underlying datasets" affects determinations of "who is and who is not a combatant." These biases can manifest in multiple ways:

Data Bias: Training datasets that reflect historical prejudices can lead AI systems to disproportionately target specific ethnic groups, age demographics, or gender categories. Experts warn that "lack of diversity in datasets may cause AI systems to single out, for example, members of distinct ethnic groups as targets, or even consider all civilian males as combatants due to encoded gender biases."

Design Bias: Human developers unconsciously embed their own prejudices into AI systems during the development phase. The creation of targeted kill lists is particularly troubling since this process depends on data inputs consistent with prevailing societal prejudices, including potentially biased assumptions about what characteristics indicate security threats.

Operational Bias: Even systems without inherent bias can develop problematic patterns through use. AI systems can reinforce their own biases through continuous learning, creating feedback loops where initial mistakes become entrenched patterns. As researchers noted, "if an AI DSS consistently identifies people of a particular gender with a particular physical appearance from a certain neighbourhood as suspected threats, it can reinforce its bias by assuming that all people matching these characteristics from that neighbourhood are 'threats actors.'"

System Brittleness and Hallucinations

AI systems demonstrate dangerous "brittleness"—an inability to adapt correctly when presented with data they haven't encountered before. This can lead to catastrophic misinterpretations. For example, if an AI system learns that enemy combatants once used a school bus for military purposes, it might subsequently classify all school buses as legitimate military targets.

"Hallucinations"—where AI systems perceive patterns or objects that don't exist—pose particularly severe risks in military contexts. When paired with unexpected changes on the battlefield, this may cause systems to "perceive patterns where there is none, prompting the targeting of innocent civilians." A 2021 United Nations Institute for Disarmament Research report highlighted how autonomous systems might lock onto trees instead of people (militarily ineffective) or worse, people instead of trees (potentially lethal to civilians).

Adversarial Attacks and Spoofing

Enemy forces can deliberately manipulate AI systems through various attack vectors. Research has shown that simple modifications—such as using tape to alter speed limit signs—can fool AI systems into dangerous behaviors. In military contexts, hostile forces could use similar techniques to:

- **Data Poisoning**: Feeding false information into AI systems during training or operation
- **Spoofing**: Creating false signatures that fool AI recognition systems
- **Adversarial Examples**: Subtly modifying inputs to cause AI systems to misclassify targets

The Arms Control Association warned in November 2024 that such "data poisoning attacks could lead to faulty early-warning systems and associated algorithms that are meant to recognize patterns or generate options. If undetected, such tampering could lead to misguided and potentially escalatory behaviors during a crisis."

Organizational Risks

Loss of Human Skills and Strategic Thinking

As military organizations become increasingly dependent on AI systems, they risk experiencing atrophy in fundamental human capabilities. The most immediate concern is automation bias—where human operators develop "unquestioning trust" in AI outputs and lose the ability to think critically about machine recommendations.

Research published in War on the Rocks in May 2024 found that military personnel who perceive AI agents as more capable than they actually are may become predisposed to "social loafing" or complacency in critical tasks such as target acquisition and intelligence gathering. This phenomenon, similar to how drivers using anti-lock braking systems drive more aggressively, could have catastrophic consequences in military contexts.

More fundamentally, over-reliance on AI risks degrading the strategic thinking capabilities that proved crucial in historical crises. The intuition and contextual understanding that allowed Stanislav Petrov to recognize a false alarm, or that enabled Kennedy's careful deliberation during the Cuban Missile Crisis, could be lost as commanders become accustomed to machine-speed decision-making.

Democratic Erosion and Accountability Gaps

The rapid deployment of AI military systems poses significant challenges to democratic oversight. With over 180 instances of AI military applications identified by the Pentagon, maintaining meaningful civilian control becomes increasingly difficult. As the Center for Strategic and International Studies noted, "no one's come up with great answers for appropriate classification of AI products," creating accountability gaps where secret AI capabilities may operate beyond public scrutiny.

The "classification by compilation" problem is particularly troubling—AI systems can piece together classified information from ostensibly unclassified sources, potentially revealing sensitive information or making decisions based on data that hasn't been properly cleared. This creates a scenario where AI systems might operate with knowledge that their human overseers don't possess, fundamentally inverting the chain of command.

Public Citizen's November 2024 report warned of an emerging "military-industrial-AI complex" that could reduce civilian control over military decision-making. When AI companies like OpenAI revised their usage guidelines in January 2024 to allow military applications previously prohibited, it highlighted how corporate decisions can directly impact national security policy.

Alliance Fragmentation and Interoperability Failures

NATO's own analysis reveals that AI systems developed by different nations "may contradict each other and slow down targeting cycle rather than speeding it up." The alliance's 2024 revised AI strategy acknowledged the need for autonomous systems to be "programmed to operate together without relying on common sense or relationships between human operators."

The technical challenge is compounded by political tensions. The United States' restrictions on AI sharing exclude 22 NATO members from "trusted countries" lists, leading to protests from allies like Poland and Israel. As these restrictions fragment the alliance's technological cooperation, they may inadvertently create vulnerabilities that adversaries could exploit.

Even more concerning is the potential for AI systems from allied nations to engage each other autonomously during joint operations. Without carefully coordinated programming and extensive testing, friendly fire incidents involving AI systems could escalate into broader conflicts between allies.

Specific Failure Modes

Nuclear Command and Control Vulnerabilities

The integration of AI into nuclear command and control systems represents the highest-stakes arena for potential failures. Despite U.S. policy maintaining "humans in the loop for nuclear weapons employment," experts warn that this policy "leaves unaddressed several dangerous ways in which AI, C3, and nuclear weapons systems could become entangled."

The fundamental problem is "entanglement"—nuclear command and control systems are inevitably connected to conventional forces that increasingly rely on AI. As a former U.S. Air Force deputy chief of staff noted, nuclear operations require "seamless integration of conventional and nuclear forces." When conventional systems that support nuclear missions become AI-enabled, the distinction between human and machine control becomes blurred.

Chatham House warned in June 2025 that false positives in AI-enhanced early warning systems "could have devastating consequences" due to the difficulty in explaining neural networks' decision-making models. The 1983 Petrov incident offers a sobering reminder: if that Soviet early warning system had been powered by AI trained on prevailing doctrine (which called for immediate retaliation to any detected attack), the machine might have recommended nuclear launch rather than questioning the alert.

Autonomous System Conflicts and Escalation Spirals

The prospect of autonomous weapons systems engaging each other without human oversight creates unprecedented escalation risks. The Centre for International Governance Innovation warned that "a deadly accident between the PLA and US military, or one of Washington's allies, involving these weapons around Taiwan or in the South China Sea, could trigger a globally destabilizing war that no one really wants."

Such incidents could escalate rapidly because of the speed of AI decision-making. While human commanders might have minutes or hours to de-escalate a confrontation, autonomous systems could complete engagement sequences in seconds. By the time human operators realize what's happening, the autonomous systems may have already committed acts that force their human commanders to respond militarily.

The risk is exacerbated by the "black box" nature of many AI systems. When an autonomous weapon engages a target, human operators may not understand why the system made that decision, making it difficult to assess whether the action was justified or if similar incidents should be expected.

Information Warfare Blowback

AI-enhanced information operations can spiral out of control in ways their creators never intended. China's deployment of "sophisticated machine learning models to create personalized propaganda" and its systematic use of "AI, big data analytics, and psychological operations to manipulate enemy perceptions" represent new forms of cognitive warfare that could have unintended consequences.

The danger lies in AI systems that continue learning and adapting during operation. An AI propaganda system designed to influence foreign audiences might eventually begin affecting domestic populations, or an AI system designed to detect enemy disinformation might develop biases that make it unable to distinguish between legitimate dissent and hostile manipulation.

Russian cyberattacks on Ukrainian AI companies developing battlefield solutions, as reported in 2024, demonstrate how AI systems themselves become targets in information warfare. When adversaries successfully attack or manipulate AI systems, the resulting failures could trigger military responses based on false information.

Mitigation Strategies

Technical Safeguards and Human Oversight

The most fundamental mitigation strategy is maintaining meaningful human control over critical decisions. However, as the Department of Defense's chief digital and AI officer noted in February 2024, "Will there always be a human in the loop for every autonomous decision? I don't think that's possible anymore." This reality requires more sophisticated approaches to human-AI integration.

NATO's 2024 AI strategy emphasizes six principles of responsible use: lawfulness, responsibility, explainability, reliability, governability, and bias mitigation. The alliance's Defense Innovation Accelerator (DIANA) network works to develop "AI certification standards" that can provide common testing and evaluation protocols across member nations.

Technical safeguards must address the full lifecycle of AI systems. The Modern War Institute recommends implementing evaluation tools prior to deployment that test AI systems against known adversarial methods. These tools should provide quantitative measures of system robustness and identify potential failure modes before systems are fielded.

Policy Frameworks and Legal Constraints

Legislative measures are beginning to emerge to constrain the most dangerous applications of military AI. The "Block Nuclear Launch by Autonomous AI Act," introduced in Congress, would prohibit nuclear weapon launches by automated systems without "meaningful human control." Similar legislation could be extended to other high-risk applications.

International cooperation on AI governance is advancing through forums like the Responsible AI in the Military Domain (REAIM) summits. Over 40 nations have joined U.S. endorsements of responsible military AI declarations, though major powers like China and Russia remain outside these frameworks.

The European Union's AI Act represents the most comprehensive regulatory approach, establishing a "human-centric, risk-based model" that contrasts with more permissive U.S. approaches. While not specifically designed for military applications, the EU framework provides a model for how democratic societies might regulate high-risk AI systems.

Institutional Adaptations and Training

Military institutions must fundamentally adapt to the realities of human-AI collaboration. The Pentagon's emphasis on ensuring "combat leaders understand basics of AI and machine learning" represents a crucial first step, but much more comprehensive changes are needed.

Training programs must go beyond technical education to include regular exercises that test human-AI teams under stress. These exercises should specifically test for automation bias, system failures, and adversarial attacks. Military personnel need to develop the skills to recognize when AI systems are providing questionable recommendations and the confidence to override machine judgments when necessary.

The Massachusetts Institute of Technology's Horizon program provides AI training for military personnel, offering workshops on the intersection of AI and military operations. Such programs need to be expanded and made mandatory for personnel working with AI-enabled systems.

Case Studies in AI Failure

The Gaza AI Targeting Controversy

Reports from 2024 revealed concerning details about AI targeting systems used in military operations. According to investigations by +972 Magazine and Local Call, AI-enabled systems like "Lavender" were reportedly used to choose human targets based on behavioral patterns with "little human oversight." These systems allegedly demonstrated several failure modes that highlight broader risks:

- Automation bias, where human operators became "cogs in a mechanised process" and lost "the ability to consider the risk of civilian harm in a meaningful way"
- Algorithmic bias that may have disproportionately targeted certain demographic groups
- Speed of operations that prevented adequate human review of targeting decisions

While specific details remain contested, the reports illustrate how AI targeting systems can fail in ways that increase rather than decrease civilian casualties.

Commercial AI System Failures with Military Implications

Civilian AI failures provide concerning previews of potential military system failures. A 2021 incident where a Cruise self-driving vehicle drove into wet concrete in San Francisco highlights the "edge case" problem that affects all AI systems. As AI expert Gary Marcus noted, "The more complicated a domain is, the more unanticipated outliers there tend to be."

Military environments are far more complex and unpredictable than civilian roads, suggesting that military AI systems will face even more numerous and dangerous edge cases. The 2017 incident where a small U.S. Army drone flew over 600 miles from Arizona to Denver after operators lost communications demonstrates how even simple autonomous systems can behave unpredictably when they encounter situations they weren't designed for.

NATO Exercise Vulnerabilities

NATO exercises have revealed significant vulnerabilities in AI system coordination. Alliance officials report that computer vision systems from different nations can contradict each other, potentially slowing down rather than speeding up military operations. During one exercise, autonomous systems from different allies engaged in conflicting behaviors that would have been catastrophic in actual combat.

These exercises also revealed how difficult it is to program autonomous systems to "operate together without relying on common sense or relationships between human operators." The implicit understanding that enables human military units to coordinate effectively cannot easily be replicated in machine systems.

Early Warning Systems

Technical Indicators of System Stress

Military organizations need robust systems for detecting when AI systems are operating outside their intended parameters. Key indicators include:

Performance Degradation: Sudden changes in system accuracy or confidence levels may indicate adversarial attacks or system failures. Monitoring systems should track these metrics continuously and alert human operators when performance falls below established thresholds.

Anomalous Behavior Patterns: AI systems that begin recommending actions inconsistent with their training or historical patterns may be experiencing failures or manipulation. These anomalies should trigger immediate human review.

Cross-System Conflicts: When multiple AI systems provide contradictory recommendations, this may indicate either system failures or situations beyond their training parameters. Such conflicts should automatically escalate to human decision-makers.

Organizational Warning Signs

Beyond technical indicators, military organizations must monitor for organizational warning signs that suggest dangerous over-reliance on AI systems:

Decreased Human Expertise: When human operators lose the ability to perform tasks without AI assistance, the organization becomes vulnerable to catastrophic failures if AI systems fail.

Automation Bias Incidents: Cases where human operators accept AI recommendations despite contradictory information suggest dangerous levels of automation bias.

Classification Gaps: When AI systems consistently operate faster than human oversight can keep pace, dangerous accountability gaps emerge.

International Monitoring Systems

The Arms Control Association has proposed international monitoring systems to track AI military developments and detect potentially destabilizing deployments. Such systems might include:

Transparency Measures: Regular reporting on military AI capabilities and deployment timelines could help prevent arms races and miscalculations.

Incident Sharing: Creating mechanisms for reporting and analyzing AI system failures could help all nations learn from mistakes and improve safety.

Crisis Communication: Establishing direct communication channels between major powers for AI-related incidents could prevent escalation when systems fail or conflict.

Governance Frameworks and International Cooperation

Multilateral Initiatives

The United Nations has taken increasing interest in military AI governance. UN Secretary-General António Guterres called in August 2024 for concluding a "legally binding instrument by 2026" to prohibit lethal autonomous weapons systems that function without human control. The UN General Assembly adopted Resolution L.56 in October 2023, entitled "Promoting International Cooperation on Peaceful Uses in the Context of International Security."

However, international consensus remains elusive. The Stop Killer Robots advocacy organization noted that while 47 states expressed support for some form of prohibition or regulation of lethal autonomous weapons systems, "many prominent countries remain opposed to or are undecided on a potential treaty regulating these weapons, including China, India, Israel, Japan, Russia, the United Kingdom, and the United States."

Bilateral Cooperation Models

The most significant bilateral development occurred in November 2024 when Chinese President Xi Jinping and U.S. President Joe Biden agreed that artificial intelligence should never be empowered to decide to launch nuclear war. This agreement, developed through five years of Track II dialogue between the Brookings Institution and Tsinghua University, provides a foundation for further cooperation.

However, broader cooperation remains limited by strategic competition. The Brookings Institution noted that "the groundwork for this excellent policy decision was laid over five years of discussions," suggesting that meaningful AI governance agreements require sustained diplomatic engagement over long periods.

Regional Approaches

The European Union has taken the most comprehensive regulatory approach through its AI Act, establishing risk-based governance for AI systems. While not specifically designed for military applications, the EU framework provides a model for democratic oversight of high-risk AI systems.

NATO's approach emphasizes interoperability and collective security through initiatives like DIANA and the €1 billion Innovation Fund. The alliance's challenge lies in balancing collective security benefits with members' desires to maintain national AI advantages.

Key Warning Development

The evidence presented in this chapter leads to an inescapable conclusion: we are rapidly approaching a point where human control over strategic decisions may be lost not through deliberate choice, but through the cumulative effect of numerous incremental AI deployments that interact in unpredictable ways.

The warning signs are clear:

1. **Speed of deployment is outpacing safety assessment**: Military AI budgets are increasing exponentially while safety frameworks lag behind
2. **Integration complexity is exceeding human comprehension**: AI systems across multiple domains are creating interaction effects that no human can fully understand
3. **International cooperation is fragmenting rather than strengthening**: Strategic competition is preventing the collective action needed to manage AI risks

The window for conscious control over this transformation is rapidly closing. As former Google CEO Eric Schmidt warned, "AI will be the most disruptive force in history," and nowhere are the stakes higher than in military applications where mistakes can trigger irreversible consequences.

The choice before us is stark: either develop robust governance frameworks for military AI now, while humans still maintain meaningful control, or risk sleepwalking into an era where machines make strategic decisions faster than humans can comprehend, much less control.

The Petrov moment of the AI age may come without warning, and unlike in 1983, there may be no human hand on the switch to prevent catastrophe. The time for action is now, before that choice is taken away from us by the very systems we created to serve us.

Verifiable References and Sources

1. **International Committee of the Red Cross** (September 4, 2024). "The risks and inefficacies of AI systems in military targeting support." https://blogs.icrc.org/law-and-policy/2024/09/04/the-risks-and-inefficacies-of-ai-systems-in-military-targeting-support/
2. **Public Citizen** (November 22, 2024). "A.I. Joe: The Dangers of Artificial Intelligence and the Military." https://www.citizen.org/article/ai-joe-report/

3. **Center for Security and Emerging Technology** (December 13, 2023). "Reducing the Risks of Artificial Intelligence for Military Decision Advantage." https://cset.georgetown.edu/publication/reducing-the-risks-of-artificial-intelligence-for-military-decision-advantage/
4. **War on the Rocks** (January 17, 2024). "Artificial Intelligence and Nuclear Stability." https://warontherocks.com/2024/01/artificial-intelligence-and-nuclear-stability/
5. **Brookings Institution** (February 28, 2025). "How unchecked AI could trigger a nuclear war." https://www.brookings.edu/articles/how-unchecked-ai-could-trigger-a-nuclear-war/
6. **War on the Rocks** (December 5, 2024). "Beyond Human-in-the-Loop: Managing AI Risks in Nuclear Command-and-Control." https://warontherocks.com/2024/12/beyond-human-in-the-loop-managing-ai-risks-in-nuclear-command-and-control/
7. **Stockholm International Peace Research Institute** (2025). "Impact of Military Artificial Intelligence on Nuclear Escalation Risk." https://www.sipri.org/publications/2025/sipri-insights-peace-and-security/impact-military-artificial-intelligence-nuclear-escalation-risk
8. **Human Rights Watch** (February 16, 2023). "Review of the 2023 US Policy on Autonomy in Weapons Systems." https://www.hrw.org/news/2023/02/14/review-2023-us-policy-autonomy-weapons-systems
9. **Modern Diplomacy** (August 22, 2024). "Publicizing the Bugs, Malfunctions, Failures, and Locations of Lethal AWS to Avoid Accidental Escalations." https://moderndiplomacy.eu/2024/08/22/publicizing-the-bugs-malfunctions-failures-and-locations-of-lethal-aws-to-avoid-accidental-escalations/
10. **Total Military Insight** (July 13, 2024). "The Role and Implications of Autonomous Weapon Systems." https://totalmilitaryinsight.com/autonomous-weapon-systems/
11. **Arms Control Association** (October 2024). "Guterres Calls for Autonomous Weapons Controls." https://www.armscontrol.org/act/2024-10/news/guterres-calls-autonomous-weapons-controls
12. **Undark** (January 1, 2025). "The Technology for Autonomous Weapons Exists. What Now?" https://undark.org/2024/11/26/unleashed-autonomous-weapons/
13. **Stop Killer Robots** (2024). "Problems with autonomous weapons." https://www.stopkillerrobots.org/stop-killer-robots/facts-about-autonomous-weapons/
14. **International Committee of the Red Cross** (October 18, 2023). "What you need to know about autonomous weapons." https://www.icrc.org/en/document/what-you-need-know-about-autonomous-weapons
15. **Popular Science** (May 17, 2021). "Autonomous weapons could make grave errors in war." https://www.popsci.com/technology/autonomous-machines-mistakes-un-institute-report/

16. **Centre for International Governance Innovation** (2024). "The United States Quietly Kick-Starts the Autonomous Weapons Era." https://www.cigionline.org/articles/the-united-states-quietly-kick-starts-the-autonomous-weapons-era/
17. **United Nations University** (January 27, 2025). "Militarization of AI Has Severe Implications for Global Security and Warfare." https://unu.edu/article/militarization-ai-has-severe-implications-global-security-and-warfare
18. **International Committee of the Red Cross** (March 14, 2024). "Falling under the radar: the problem of algorithmic bias and military applications of AI." https://blogs.icrc.org/law-and-policy/2024/03/14/falling-under-the-radar-the-problem-of-algorithmic-bias-and-military-applications-of-ai/
19. **International Committee of the Red Cross** (September 3, 2024). "The problem of algorithmic bias in AI-based military decision support systems." https://blogs.icrc.org/law-and-policy/2024/09/03/the-problem-of-algorithmic-bias-in-ai-based-military-decision-support-systems/
20. **Fair Observer** (September 24, 2024). "We Need to Reduce Killer Military AIs Bias Immediately." https://www.fairobserver.com/more/science/we-need-to-reduce-killer-military-ais-bias-immediately/
21. **Lieber Institute West Point** (September 6, 2024). "Responsible AI Symposium – Legal Implications of Bias Mitigation." https://lieber.westpoint.edu/legal-implications-bias-mitigation/
22. **IE Insights** (October 3, 2024). "When AI Meets the Laws of War." https://www.ie.edu/insights/articles/when-ai-meets-the-laws-of-war/
23. **Arms Control Association** (November 12, 2024). "Beyond a Human 'In the Loop': Strategic Stability and Artificial Intelligence." https://www.armscontrol.org/issue-briefs/2024-011/beyond-the-loop
24. **War on the Rocks** (May 29, 2024). "Is there a Human in the Machine? AI and Future Warfare." https://warontherocks.com/2024/05/is-there-a-human-in-the-machine-ai-and-future-warfare/
25. **Modern War Institute** (January 25, 2022). "Artificial Intelligence, Real Risks: Understanding—and Mitigating—Vulnerabilities in the Military Use of AI." https://mwi.westpoint.edu/artificial-intelligence-real-risks-understanding-and-mitigating-vulnerabilities-in-the-military-use-of-ai/
26. **International Committee of the Red Cross** (August 29, 2024). "Artificial intelligence in military decision-making: supporting humans, not replacing them." https://blogs.icrc.org/law-and-policy/2024/08/29/artificial-intelligence-in-military-decision-making-supporting-humans-not-replacing-them/

27. **Washington Times** (February 26, 2024). "Defense Department AI chief: Keeping humans in the loop not always possible anymore." https://www.washingtontimes.com/news/2024/feb/23/defense-department-ai-chief-keeping-human-in-loop-/
28. **Sentient Digital, Inc.** (June 20, 2024). "Military Applications of AI in 2024." https://sdi.ai/blog/the-most-useful-military-applications-of-ai/
29. **NATO Review** (April 12, 2024). "Should artificial intelligence be banned from nuclear weapons systems?" https://www.nato.int/docu/review/articles/2024/04/12/should-artificial-intelligence-be-banned-from-nuclear-weapons-systems/index.html
30. **Chatham House** (June 9, 2025). "What happens if AI goes nuclear?" https://www.chathamhouse.org/publications/the-world-today/2025-06/what-happens-if-ai-goes-nuclear
31. **RAND Corporation** (April 24, 2018). "How Artificial Intelligence Could Increase the Risk of Nuclear War." https://www.rand.org/pubs/articles/2018/how-artificial-intelligence-could-increase-the-risk.html
32. **Carnegie Endowment for International Peace** (2024). "Governing Military AI Amid a Geopolitical Minefield." https://carnegieendowment.org/research/2024/07/governing-military-ai-amid-a-geopolitical-minefield

Chapter 15: Toward a New Strategic Equilibrium

Managing the AI Transformation Responsibly

Current Reality of Human-AI Strategic Collaboration

The future of strategic decision-making is not a distant possibility—it is unfolding today across military organizations worldwide. In 2025, the U.S. Army created new artificial intelligence-focused occupational specialties (49B for enlisted personnel) and warrant officer positions, recognizing that effective AI integration requires dedicated expertise embedded across formations. This represents the first formalization of new career fields in over a year, signaling the military's commitment to human-AI integration as a core capability rather than an experimental add-on.

The Pentagon's current approach to AI integration exemplifies the careful balance required between machine efficiency and human oversight. As Dr. Radha Plumb, the Pentagon's Chief Digital and AI Officer, emphasizes: "As a matter of both reliability and ethics, we'll always have humans involved in the decision to employ force." This framework of human-machine collaboration ensures that while AI systems accelerate the "kill chain" process of identifying, tracking, and addressing threats, commanders maintain authority over critical decisions.

Real-world applications demonstrate this integration in action. The Pentagon's Thunderforge project, developed by Scale AI in partnership with Microsoft and Anduril, represents operational AI deployment for strategic planning. The system helps commanders analyze multiple scenarios through AI-assisted war-gaming while maintaining human oversight throughout the process. As Pentagon officials explain, "decision cycles are measured in seconds, not days," but humans continue to make final strategic determinations based on AI-generated options and analysis.

The Marine Corps has pioneered practical applications through its School of Advanced Warfighting, using large language models to connect and visualize different layers of information—from strategic-level economic relationships to focused country-specific dynamics. These systems enable planners to perform in hours what previously took days, allowing human decision-makers to concentrate on the most relevant insights derived from massive datasets.

Lessons Learned: National Approaches to AI Strategic Transformation

The past three years have provided crucial insights into how different strategic cultures approach AI integration, revealing distinct patterns and their relative strengths and limitations.

United States: Democratic Innovation Under Pressure

The U.S. model demonstrates both the advantages and challenges of democratic oversight combined with competitive innovation. With an $1.8 billion AI budget for fiscal 2025 and over 180 documented AI military applications, America's approach balances rapid technological adoption with accountability mechanisms.

Key strengths include robust public-private partnerships enabling rapid innovation cycles and the integration of commercial AI advances into military applications. The recent awards of $200 million contracts to Google, xAI, Anthropic, and OpenAI for developing "agentic AI workflows" illustrate this approach's ability to leverage cutting-edge commercial technology for national security purposes.

However, this model faces significant challenges. The Pentagon's Chief Digital and AI Officer Doug Matty notes the difficulty of "leveraging commercially available solutions into an integrated capabilities approach" while maintaining security and oversight. Classification challenges arise where "no one's come up with great answers for appropriate classification of AI products," particularly when AI systems can piece together classified information from unclassified sources.

The democratic oversight imperative creates tension with the speed required for effective AI integration. Congressional efforts like the "Block Nuclear Launch by Autonomous AI Act" reflect legitimate concerns about maintaining human control while potentially constraining operational flexibility.

NATO Alliance: Collaborative Standards Development

NATO's revised AI strategy, released in July 2024, represents the most comprehensive multilateral approach to military AI governance. Building on the alliance's 2021 foundation, the updated strategy addresses generative AI capabilities and establishes six Principles of Responsible Use: Lawfulness, Responsibility and Accountability, Explainability and Traceability, Reliability, Governability, and Bias Mitigation.

The alliance's approach emphasizes interoperability as a core challenge, recognizing that "future high-speed conflicts require sharing large volumes of AI-parsed data to connect sensors to shooters." NATO's response includes developing "Alliance-wide AI Testing, Evaluation, Verification & Validation (TEV&V) landscape" utilizing the Defence Innovation Accelerator (DIANA) network.

The €1 billion NATO Innovation Fund represents the "world's first multi-sovereign venture capital fund," demonstrating how alliances can pool resources for AI development while maintaining individual national advantages. This model provides a template for managing competitive cooperation in AI development.

However, alliance coordination faces inherent tensions. As NATO documents acknowledge, "AI systems developed by different nations may contradict each other and slow down targeting cycle rather than speeding it up." The challenge of programming autonomous systems to "operate together without relying on common sense or relationships between human operators" remains significant.

International Governance: Emerging Frameworks

Multiple international initiatives are attempting to establish global norms for military AI use. The Political Declaration on Responsible Military Use of Artificial Intelligence and Autonomy, launched by the United States in 2023, has gained endorsement from over 60 countries, including most NATO members and key partners.

The UN Security Council has increasingly focused on AI governance, with Secretary-General António Guterres warning that "artificial intelligence is outpacing human ability to govern it." Recent UN initiatives include Resolution 79/239 requesting views on "opportunities and challenges posed to international peace and security by the application of artificial intelligence in the military domain."

The International Committee of the Red Cross has submitted comprehensive recommendations emphasizing that under International Humanitarian Law, "parties to armed conflict – ultimately, human beings – who are responsible for complying with IHL and who must be held accountable for violations." This reinforces the principle that human accountability cannot be delegated to machines.

Framework for Managing AI Strategic Transformation

Based on current developments and expert analysis, five core principles emerge as essential for successful AI strategic transformation:

1. Deliberate Cultural Evolution

Success requires conscious strategic culture design rather than allowing technological capabilities to drive organizational change. The U.S. Army's creation of dedicated AI career fields exemplifies this approach, ensuring that human-AI integration becomes embedded in military culture through professional development and career progression.

As highlighted by defense analysts, effective AI integration requires "combat leaders understand basics of AI and machine learning." This necessitates comprehensive education reform, from basic military training through senior leadership development. The Army's emphasis on developing "Human-Machine Integrated Formations" reflects understanding that cultural adaptation must parallel technological capability.

2. Human-AI Integration Models

Current practice reveals three primary integration approaches:

Human-in-the-Loop: Direct human involvement in each decision cycle, exemplified by current U.S. nuclear policy maintaining "humans in the loop for nuclear weapons employment." This model ensures maximum human control but potentially limits AI's speed advantages.

Human-on-the-Loop: Human oversight of AI operations with intervention capability, demonstrated in the Pentagon's Thunderforge system where humans oversee AI-generated strategic planning options. This balances speed with control.

Human-out-of-the-Loop: Autonomous AI operation in defined parameters, currently limited to defensive systems like Close-In Weapons Systems (CIWS). Pentagon policy explicitly states this "does not apply to cyber" domains, indicating selective application based on risk assessment.

3. Democratic Oversight and Accountability

Maintaining democratic control over AI strategic systems requires new approaches to oversight that account for technological complexity and operational speed requirements. Current U.S. efforts include Congressional involvement in AI policy development and public accountability for military AI applications.

The challenge lies in balancing necessary oversight with operational effectiveness. As military leaders note, traditional oversight mechanisms designed for deliberate planning cycles may not suit AI-enabled rapid decision-making. Solutions include:

- Pre-approved operational parameters for AI systems
- Real-time monitoring and logging of AI decision-making
- Regular audits of AI system performance and bias
- Clear rules of engagement for AI-enabled operations

4. Alliance Coordination and Interoperability

NATO's approach provides a model for managing AI coordination among allies while preserving national advantages. Key elements include:

Common Standards: Development of shared technical specifications enabling AI system interoperability without requiring identical systems. NATO's DIANA network facilitates this through coordinated testing and certification.

Data Sharing Protocols: Establishment of secure frameworks for sharing AI training data and operational intelligence. NATO's Digital Transformation Implementation Strategy emphasizes "data sharing as a shared responsibility" across the alliance.

Crisis Management Procedures: Development of protocols for managing AI system conflicts or failures during joint operations. This includes communication procedures and fallback mechanisms when AI systems malfunction.

5. Continuous Adaptation and Learning

AI's rapid evolution necessitates organizational cultures capable of continuous learning and adaptation. As defense experts note, AI systems require updates "every six to 12 weeks" to remain effective, demanding organizational agility beyond traditional military culture.

This requires:

- Rapid experimentation and testing protocols
- Failure-tolerant organizational cultures
- Continuous training and education systems
- Regular assessment and adjustment of AI integration strategies

Policy Recommendations for Implementation

For Military Leaders

Current best practices suggest several immediate actions for military leadership:

Invest in Human Development: The Pentagon's emphasis on ensuring "combat leaders understand AI basics" reflects recognition that successful integration depends on human capability as much as technological sophistication. This requires comprehensive education reform across all levels of military service.

Develop Human-AI Teaming Doctrines: Military organizations must establish clear frameworks for human-AI collaboration in operational contexts. The Marine Corps' experimentation with large language models in war-gaming provides a model for practical doctrine development.

Maintain Human Decision-Making Skills: Avoiding over-dependence on AI systems requires deliberate preservation of human strategic thinking capabilities. This includes regular exercises without AI support and emphasis on human creativity and adaptability.

Create Red Team Exercises: Testing AI system vulnerabilities through adversarial exercises helps identify potential failure modes before operational deployment. Current Pentagon initiatives include testing AI systems against human manipulation and deception.

For Policymakers

Establish Clear Legal Frameworks: Current efforts like the Congressional "Block Nuclear Launch by Autonomous AI Act" demonstrate the need for explicit legal boundaries on AI military applications. These frameworks must balance operational flexibility with democratic oversight.

Invest in Education and Workforce Development: The broader challenge of AI strategic transformation requires national investment in AI education and technical expertise. This includes support for STEM education and programs to attract AI talent to government service.

Foster International Cooperation: Participation in initiatives like NATO's AI strategy and the UN's governance frameworks helps establish global norms while protecting national interests. The U.S. leadership in the Political Declaration on Responsible Military Use demonstrates effective diplomatic engagement.

Maintain Ethical Standards: Democratic values require explicit consideration of ethical implications in AI military applications. This includes transparency in AI development, accountability for AI decisions, and protection of civilian populations.

For Technologists

Design for Human Oversight: AI systems must include capabilities for human intervention and explanation. NATO's emphasis on "explainability and traceability" in AI systems reflects operational requirements for human understanding and control.

Consider Operational Context: Military AI development must account for the contested, degraded environments where these systems will operate. This includes designing for resilience against cyber attacks and electronic warfare.

Collaborate with Users: Effective AI systems require close collaboration between developers and military users to ensure practical utility. Current Pentagon contracts emphasize this partnership approach.

Prioritize Reliability and Testing: Military applications demand higher reliability standards than commercial AI systems. This requires extensive testing under realistic operational conditions and robust quality assurance processes.

For Alliance Partners

Develop Common Standards: NATO's DIANA network provides a model for collaborative standard development that maintains interoperability while preserving national technological advantages.

Share Best Practices: Regular exchanges of lessons learned help all alliance members benefit from collective experience. This includes sharing both successes and failures in AI integration.

Coordinate Research and Development: Joint AI research initiatives can pool resources while avoiding duplication of effort. The NATO Innovation Fund exemplifies this approach.

Establish Crisis Management Protocols: Procedures for managing AI system conflicts or failures during joint operations help maintain alliance cohesion during crises.

Path to AI-Enhanced Strategic Stability

Near-Term (2025-2030): Foundation Building

Current trends suggest several immediate priorities for establishing stable AI strategic governance:

International Norms Development: The momentum behind initiatives like the Political Declaration on Responsible Military Use and NATO's AI strategy provides opportunity for establishing global norms before AI military applications become more widespread.

Technical Standards Establishment: NATO's work on AI certification standards through the DIANA network offers a model for developing technical frameworks that ensure reliable and interoperable AI systems.

Human-AI Integration Capability Building: Military organizations must invest immediately in developing human-AI collaboration capabilities. The U.S. Army's creation of dedicated AI career fields provides a template for institutionalizing this capability.

Alliance Cooperation Framework Development: NATO's revised AI strategy and the U.S.-led Political Declaration demonstrate the potential for multilateral cooperation on AI governance. Expanding these frameworks to include more partners enhances global stability.

Medium-Term (2030-2035): Operational Integration

Successful foundation building enables more ambitious AI integration while maintaining stability:

Stable Deterrence with AI-Enhanced Forces: As AI capabilities mature, maintaining strategic stability requires ensuring that AI enhancement strengthens rather than undermines deterrence relationships. This includes clear communication of AI capabilities and limitations to potential adversaries.

Robust Crisis Management Systems: AI's speed advantages require new approaches to crisis management that can operate at machine speed while preserving human decision-making authority. This includes automated communication systems and pre-negotiated protocols for AI system interactions.

AI Arms Control Mechanisms: As AI military applications become more sophisticated, traditional arms control approaches may require adaptation. Verification mechanisms for AI capabilities present unique challenges requiring innovative solutions.

Next-Generation Strategic Culture Development: Military organizations must develop strategic cultures that seamlessly integrate human and artificial intelligence. This requires generational change in military leadership and fundamental rethinking of strategic planning processes.

Long-Term (2035-2050): Equilibrium Management

The ultimate goal involves managing AI strategic transformation in ways that enhance rather than undermine global stability:

Post-Human Strategic Environment Management: As AI capabilities approach or exceed human cognitive abilities in specific domains, maintaining human agency in strategic decision-making becomes increasingly challenging. Successful management requires clear boundaries and robust safeguards.

Quantum-AI Integration: The convergence of quantum computing and artificial intelligence may create capabilities that fundamentally alter strategic relationships. Preparing for this transition requires early research and policy development.

Human Agency Preservation: Maintaining meaningful human control over strategic decisions remains essential for democratic governance and ethical conduct of warfare. This requires continued evolution of human-AI integration models.

Multi-Polar AI Strategic Balance: A stable global order requires balancing AI capabilities among major powers while preventing AI advantages from creating instability. This necessitates continued international cooperation and norm development.

International Cooperation Models

Multilateral Frameworks

UN-Based Governance: Recent UN initiatives, including Security Council discussions and the Secretary-General's reports on AI in the military domain, provide global legitimacy for AI governance efforts. However, these mechanisms face challenges in achieving consensus among competing powers.

Regional Cooperation: NATO's AI strategy demonstrates how regional organizations can develop effective governance frameworks among like-minded partners. Similar approaches in other regions could complement global efforts.

Issue-Specific Partnerships: Cooperation on specific AI challenges, such as humanitarian applications or arms control verification, may prove more achievable than comprehensive agreements.

Bilateral Cooperation

Strategic Dialogue: Regular AI strategic dialogues between major powers, similar to traditional arms control discussions, could help manage competitive dynamics while reducing risks of miscalculation.

Technical Cooperation: Collaboration on AI safety research and testing protocols benefits all parties by improving system reliability and reducing shared risks.

Confidence-Building Measures: Transparency initiatives regarding AI military capabilities could reduce uncertainty and prevent arms race dynamics.

Track-Two Initiatives

Academic and Research Cooperation: Continued collaboration among AI researchers on safety and governance questions helps develop technical solutions to policy challenges.

Military-to-Military Exchanges: Professional exchanges among military AI practitioners can build understanding and trust while sharing best practices.

Private Sector Engagement: Given the private sector's leading role in AI development, engaging technology companies in governance discussions helps align commercial development with security needs.

Success Metrics and Indicators

Measuring progress in AI strategic transformation requires new approaches to assessment that account for both technological capability and strategic stability:

Capability Metrics

Integration Effectiveness: Measures of how successfully AI systems enhance rather than complicate military decision-making. The Pentagon's emphasis on "decision advantage" provides one framework for assessment.

Reliability and Performance: Technical metrics for AI system accuracy, reliability, and performance under operational conditions. The 96% accuracy achieved by the Pentagon's Scylla threat detection system provides a benchmark for evaluation.

Interoperability: Assessment of how effectively AI systems from different organizations and nations can work together. NATO's emphasis on alliance-wide compatibility provides standards for measurement.

Stability Indicators

Crisis Management Effectiveness: Evaluation of how AI-enhanced forces perform during international crises, including whether AI systems contribute to escalation or de-escalation.

Alliance Cohesion: Assessment of whether AI integration strengthens or weakens alliance relationships. NATO's continued cooperation on AI development provides positive indicators.

Strategic Predictability: Measures of whether AI capabilities contribute to strategic stability through predictable behavior or create instability through unpredictable effects.

Governance Effectiveness

Democratic Oversight: Evaluation of whether democratic institutions successfully maintain control over AI military applications. Congressional engagement in AI policy provides one measure of democratic involvement.

International Cooperation: Assessment of progress in developing international norms and cooperation mechanisms for AI governance. The growing endorsement of responsible use declarations indicates positive trends.

Ethical Compliance: Monitoring of whether AI military applications comply with established ethical standards and international humanitarian law.

The Imperative for Action

The window for conscious strategic culture design is rapidly closing. Current trends in AI development suggest that the next five years will determine whether AI integration enhances or undermines strategic stability. Several factors create urgency for immediate action:

Technological Acceleration: AI capabilities continue advancing at unprecedented pace, with new developments potentially disrupting current assumptions about human-AI integration. The rapid evolution of large language models and autonomous systems compresses decision-making timelines for policy development.

Competitive Pressures: International competition in AI development creates incentives for rapid deployment potentially at the expense of safety and governance considerations. The current "AI arms race" dynamic requires proactive governance to prevent destabilizing outcomes.

Operational Reality: AI systems are already being deployed in operational contexts, as demonstrated by recent conflicts and current military applications. Policy and governance frameworks must catch up to operational reality rather than remaining theoretical exercises.

Generational Change: Military and political leaders who will make future AI strategic decisions are entering service in an AI-enabled environment. Their assumptions about human-AI integration will shape strategic culture for decades to come.

Call to Action

Strategic leaders across all levels—military, political, and technological—must embrace the responsibility of consciously designing strategic cultures for the AI age. This requires several immediate commitments:

Military Leadership: Investment in human-AI integration education and training, development of robust testing and evaluation protocols, and maintenance of human strategic thinking capabilities alongside AI enhancement.

Political Leadership: Establishment of clear legal and policy frameworks for AI military applications, investment in national AI expertise and education, and engagement in international cooperation on AI governance.

Technological Leadership: Design of AI systems that preserve human oversight and control, collaboration with military users on operational requirements, and commitment to responsible development practices.

Alliance Partners: Coordination on technical standards and operational procedures, sharing of best practices and lessons learned, and development of cooperative governance mechanisms.

International Community: Engagement in multilateral governance initiatives, development of global norms for responsible AI use, and creation of conflict prevention and management mechanisms for the AI age.

The challenges are significant, but the opportunities for enhanced security and human flourishing are equally substantial. Artificial intelligence represents both humanity's greatest strategic opportunity and its most significant strategic challenge. Success requires not just technological advancement but the wisdom to integrate these capabilities in ways that serve human values and global stability.

The strategic choices made in the next few years will echo through decades of international relations. We have the opportunity to shape AI integration consciously and responsibly, ensuring that these powerful technologies enhance rather than undermine human agency in strategic decision-making. The alternative—allowing technological capabilities to drive strategic change without conscious guidance—risks outcomes that serve neither security nor human welfare.

The transformation of strategic cultures through artificial intelligence is not a distant possibility—it is our current reality. The question is not whether we will integrate AI into strategic affairs, but whether we will do so wisely. That choice remains ours to make, but the window for making it is rapidly closing.

Verifiable References and Sources

1. U.S. Army Creates New AI-Focused Military Occupational Specialty (49B), Military.com, July 2, 2025: https://www.military.com/daily-news/2025/07/02/army-creating-new-

artificial-intelligence-focused-occupational-specialty-and-officer-field.html

2. Pentagon Awards Multiple Companies $200M Contracts for AI Tools, Nextgov/FCW, July 15, 2025: https://www.nextgov.com/acquisition/2025/07/pentagon-awards-multiple-companies-200m-contracts-ai-tools/406698/
3. Pentagon: AI Accelerates Decision-Making in Military Operations, Technology.org, January 20, 2025: https://www.technology.org/2025/01/20/pentagon-with-ai-we-accelerate-decision-making-in-lethal-military-operations/
4. Pentagon AI "Kill Chain" Acceleration, TechCrunch, January 22, 2025: https://techcrunch.com/2025/01/19/the-pentagon-says-ai-is-speeding-up-its-kill-chain/
5. NATO Releases Revised AI Strategy, NATO Press Release, July 10, 2024: https://www.nato.int/cps/en/natohq/news_227234.htm
6. NATO's Revised Artificial Intelligence Strategy Summary, July 10, 2024: https://www.nato.int/cps/en/natohq/official_texts_227237.htm
7. NATO Updates AI Strategy for the Age of ChatGPT, Breaking Defense, July 10, 2024: https://breakingdefense.com/2024/07/nato-updates-ai-strategy-for-the-age-of-chatgpt/
8. Political Declaration on Responsible Military Use of AI, U.S. State Department, January 16, 2025: https://www.state.gov/bureau-of-arms-control-deterrence-and-stability/political-declaration-on-responsible-military-use-of-artificial-intelligence-and-autonomy
9. UN Security Council Debates AI in Conflicts, UN Press Release, December 2024: https://press.un.org/en/2024/sc15946.doc.htm
10. OpenAI for Government Launches with $200M Pentagon Win, Breaking Defense, June 17, 2025: https://breakingdefense.com/2025/06/openai-for-government-launches-with-200m-win-from-pentagon-cdao/
11. Pentagon Signs AI Deal for Military Planning, Washington Post, March 5, 2025: https://www.washingtonpost.com/technology/2025/03/05/pentagon-ai-military-scale/
12. How Modern Militaries are Leveraging AI, Atlantic Council, August 17, 2023: https://www.atlanticcouncil.org/in-depth-research-reports/report/how-modern-militaries-are-leveraging-ai/
13. Military Applications of AI in 2024, Sentient Digital, June 20, 2024: https://sdi.ai/blog/the-most-useful-military-applications-of-ai/
14. AI in Military Decision-Making, ICRC Blog, August 29, 2024: https://blogs.icrc.org/law-and-policy/2024/08/29/artificial-intelligence-in-military-decision-making-supporting-humans-not-replacing-them/
15. Military AI Governance Challenges, ICRC Blog, December 12, 2024: https://blogs.icrc.org/law-and-policy/2024/12/12/the-im-possibility-of-responsible-military-ai-governance/

16. Pentagon Ready to Let AI Make Some Decisions, National Defense Magazine, April 5, 2025: https://www.nationaldefensemagazine.org/articles/2025/4/5/pentagon-ready-to-let-ai-make-some-decision
17. The U.S. Army, AI, and Mission Command, War on the Rocks, March 4, 2025: https://warontherocks.com/2025/03/the-u-s-army-artificial-intelligence-and-mission-command/
18. Military AI Applications in 2025, Cevians, February 3, 2025: https://www.cevians.com/military-applications-of-ai-in-2025/
19. NATO Parliamentary Assembly Report on NATO and AI, November 2024: https://www.nato-pa.int/document/2024-nato-and-ai-report-clement-058-stc
20. NATO Digital Transformation Implementation Strategy, October 17, 2024: https://www.nato.int/cps/en/natohq/official_texts_229801.htm
21. AI Leaders Shaping US Government & Military, Executive Gov, June 2025: https://executivegov.com/article/ai-leaders-us-government-military-dod-anthropic-cdao-openai-cia
22. UN AI for Good Global Summit 2025, UN News, July 9, 2025: https://news.un.org/en/story/2025/07/1165346
23. ICRC Recommendations on AI in Military Domain, ICRC, April 17, 2025: https://www.icrc.org/en/article/artificial-intelligence-military-domain-icrc-submits-recommendations-un-secretary-general
24. Global Military AI Governance Analysis, European Leadership Network: https://europeanleadershipnetwork.org/commentary/what-does-global-military-ai-governance-need/
25. Future of AI in Military Operations, Latent AI, March 28, 2025: https://latentai.com/blog/the-future-of-ai-in-military-operations-and-strategy/

CONCLUSION: THE STRATEGIC CHOICE BEFORE US

The Future of Human Agency in Strategic Decision-Making

The transformation of strategic cultures through artificial intelligence is not a distant possibility—it is happening now. From the Pentagon's 685 unclassified military AI projects funded through research and development accounts to China's commitment to achieving "intelligentized warfare" by 2027, from Ukraine's deployment of AI-enabled drones achieving 80% hit rates to NATO's revised AI strategy encompassing 32 member nations, the evidence is clear: we are living through a revolution in how nations think about and conduct strategic competition.

The Scale of Transformation

The numbers alone tell a compelling story of unprecedented change. Global military AI spending has surged from $9.67 billion in 2024 to a projected $61.08 billion by 2034, representing a compound annual growth rate of 20.57 percent. The United States has requested $1.8 billion for AI initiatives in fiscal 2025, while global defense spending reached $2,718 billion in 2024—a 9.4 percent increase and the steepest year-on-year rise since the end of the Cold War.

This financial commitment reflects a deeper strategic recognition. As documented throughout this book, artificial intelligence is not merely changing the tools of strategy but fundamentally reshaping the cultures that guide strategic thinking. The nations that recognize this transformation and consciously adapt their strategic cultures accordingly will shape the next century of international relations. Those that fail to adapt risk strategic irrelevance in an increasingly AI-driven security environment.

Evidence from the Battlefield Laboratory

Ukraine has emerged as what officials describe as a "war lab for the future," providing real-world evidence of AI's transformative impact on strategic culture. Ukrainian forces produced approximately 2 million drones in 2024, with 10,000 featuring AI-enhanced autonomous capabilities. These systems have achieved remarkable operational success, with AI-guided drones consistently hitting targets at rates exceeding 80 percent while dramatically reducing training requirements from months to mere minutes.

The strategic implications extend far beyond tactical effectiveness. As one Ukrainian drone pilot from the 23rd Mechanized Brigade observed, "in the near future, heavy armor may be unable to get within 10 kilometers of the front lines, as autonomous drones are likely to patrol the skies, systematically targeting and destroying any approaching heavy armor." This represents a fundamental shift in strategic thinking—from platform-centric to algorithm-centric warfare concepts.

Russian responses demonstrate adaptive strategic culture in action. Moscow approved a 30 percent increase in military spending for 2025, with President Vladimir Putin viewing AI as "critical for reducing a growing capability gap with the West." Russia announced a new 10-year defense strategy featuring dedicated focus on artificial intelligence, aiming to increase drone production nearly 10-fold to approximately 1.4 million units annually.

Great Power Strategic Culture Evolution

The American Democratic Innovation Model

The United States exemplifies the challenge of adapting democratic strategic culture to AI realities. The Pentagon oversees more than 685 AI-related projects, with officials acknowledging that maintaining democratic oversight over rapidly evolving AI capabilities presents unprecedented governance challenges. As one defense official noted, "no one's come up with great answers for appropriate classification of AI products," highlighting the tension between transparency and security in democratic systems.

The U.S. approach emphasizes maintaining "humans in the loop" for critical decisions while leveraging AI speed and analytical capabilities. This reflects deep-rooted American strategic culture values prioritizing individual accountability and democratic oversight, even as military leaders recognize the operational advantages of increased automation.

China's Systematic Intelligentization Drive

China's strategic culture transformation represents perhaps the most comprehensive national approach to AI integration. Chinese Communist Party General Secretary Xi Jinping's call for the People's Liberation Army to "accelerate the integrated development of mechanization, informatization, and intelligentization" by 2027 reflects systematic thinking about technological-cultural evolution.

Chinese strategists conceptualize "intelligentized warfare" as centered on the "cognitive space where effective decision-making provides decisive advantage." This represents a fundamental shift from kinetic to cognitive dominance, with AI enabling what Chinese doctrine describes as "systematic use of AI, big data analytics, and psychological operations to manipulate enemy perceptions."

The scale of Chinese commitment is evident in their investment across 57 of 64 critical technologies where China now dominates globally. The authoritarian advantage in rapid AI integration, enabled by centralized decision-making structures, demonstrates how political systems shape technological adoption patterns in strategic culture.

Russian Asymmetric AI Strategy

Russia's approach illustrates how resource constraints drive innovative strategic culture adaptation. Using AI to "offset asymmetry in conventional power due to extensive personnel and material losses in Ukraine," Russian forces have developed electronic warfare-proof autonomous systems that "process electronic threats and operate independently."

This survival-driven innovation demonstrates strategic culture's capacity for rapid evolution under existential pressure. Russian military theorists now speak of transitioning from the "guerrilla era of ad hoc drone warfare" to institutionalized autonomous systems branches, showing organizational culture adapting to technological realities.

Alliance Dynamics and Interoperability Challenges

NATO's July 2024 revised AI strategy reveals both the promise and peril of collective strategic culture transformation. The alliance's commitment to six Principles of Responsible Use—Lawfulness, Responsibility and Accountability, Explainability and Traceability, Reliability, Governability, and Bias Mitigation—represents an attempt to harmonize diverse national strategic cultures around shared AI governance norms.

However, significant challenges remain. As NATO's own analysis acknowledges, "AI systems developed by different nations may contradict each other and slow down targeting cycles rather than speeding them up." The requirement for autonomous systems to "operate together without relying on common sense or relationships between human operators" highlights the complexity of algorithmic interoperability.

The alliance's €1 billion NATO Innovation Fund and Defense Innovation Accelerator for the North Atlantic (DIANA) represent institutional adaptations to accelerate AI adoption while maintaining collective security benefits. Yet tensions persist between national AI advantages and alliance cooperation, with 22 NATO members excluded from U.S. "trusted countries" lists affecting technology sharing.

The Governance Challenge

International efforts to regulate lethal autonomous weapons systems provide insight into global strategic culture evolution. The December 2024 UN General Assembly resolution on autonomous weapons, supported by 166 nations with only 3 opposed, demonstrates unprecedented international consensus on the need for AI governance frameworks.

The momentum toward concluding "a legally binding instrument by 2026" regulating lethal autonomous weapons reflects growing recognition that technological development is outpacing governance mechanisms. UN Secretary-General António Guterres and International Committee of the Red Cross President Mirjana Spoljaric's joint call for specific prohibitions and restrictions acknowledges that "without specific rules, too much will be left to varied interpretations by states."

This governance urgency stems from accelerating deployment realities. Ukrainian forces have deployed AI-powered "mother drones" capable of delivering strike drones up to 300 kilometers behind enemy lines, with AI systems that "independently identify and select targets" without GPS reliance. Such capabilities represent crossing technological thresholds that existing international law was not designed to address.

Emerging Strategic Cultures: The Next Generation

The transformation we are witnessing extends beyond current capabilities to fundamental changes in strategic thinking. Military leaders who have grown up with AI are making fundamentally different strategic assumptions than their predecessors. This generational shift toward "AI-native strategic thinking" represents perhaps the most profound long-term implication of current developments.

Consider the evolution in training requirements: Ukrainian drone operators now require just 30 minutes of instruction to achieve effectiveness with AI-enhanced systems, compared to months of traditional training. This compression of expertise development timelines suggests strategic cultures must adapt to much faster personnel rotation and capability generation cycles.

The integration of quantum computing with AI presents additional discontinuous possibilities. Chinese investments in quantum computing for "leapfrogging scientific breakthroughs" could enable computational capabilities that fundamentally alter strategic power balances. As Chinese researchers note, quantum-AI integration offers potential for "increases in calculating speed that are impossible to imagine."

The Democratic Imperative

Democratic nations face a particular challenge in this transformation. Authoritarian systems' demonstrated advantages in rapid AI integration create pressure to sacrifice democratic oversight for competitive effectiveness. However, as this book has documented, the most successful strategic culture adaptations preserve core values while transforming capabilities.

The key lies in conscious strategic culture design rather than reactive adaptation. Democratic strategic cultures must proactively develop frameworks that maintain human agency while leveraging machine intelligence. This requires institutional innovations that can operate at AI speed while preserving accountability mechanisms.

Risks and Mitigation Strategies

The risks of poorly managed AI strategic culture transformation are profound. Historical precedent suggests that periods of rapid technological change create instability until new equilibria emerge. The current transformation involves uniquely destabilizing elements: compression of decision timelines, potential for autonomous escalation, and erosion of traditional deterrence logic.

Technical vulnerabilities compound these risks. AI systems' susceptibility to adversarial manipulation, algorithmic bias in targeting decisions, and potential for cascading failures across interconnected systems create new categories of strategic risk. As one expert warned, "what is the training data set we have for nuclear war?" highlighting the challenge of deploying AI systems in domains where failure could be catastrophic.

Mitigation requires multilayered approaches. Technical safeguards including "human in the loop" requirements for critical decisions must be combined with policy frameworks establishing clear legal boundaries. International cooperation mechanisms must evolve to address AI-specific challenges while maintaining deterrence stability.

The Path Forward: Principles for Success

Based on the analysis throughout this book, several principles emerge for successful AI strategic culture transformation:

1. Deliberate Cultural Evolution: Nations must consciously design their strategic culture evolution rather than allowing accidental development. This requires understanding how AI capabilities interact with existing cultural patterns and deliberately shaping integration processes.

2. Human-AI Integration Balance: The most successful approaches preserve human judgment while leveraging machine capabilities. This is not simply a technical challenge but a cultural one, requiring institutions that can operate effectively across human and machine timescales.

3. Democratic Oversight Mechanisms: Maintaining accountability for AI strategic systems requires developing new forms of oversight that can function at AI speed while preserving democratic values. This may require innovative approaches to classification, congressional briefing, and public accountability.

4. Alliance Coordination Frameworks: Developing "autonomous systems programmed to operate together" requires unprecedented levels of international technical cooperation combined with shared governance frameworks. Success requires treating interoperability as a cultural as well as technical challenge.

5. Continuous Adaptation Capabilities: Building strategic cultures capable of rapid learning and change while maintaining core identity represents perhaps the greatest challenge. This requires institutional designs that can evolve quickly without losing essential characteristics.

The Strategic Choice

The choice before us is not whether to embrace AI in strategic affairs—that decision has already been made by technological and competitive pressures. The choice is whether we will shape this transformation consciously and responsibly, or whether we will allow it to shape us.

The window for conscious strategic culture design is closing rapidly. As technologies advance at unprecedented pace, nations must act now to establish frameworks for AI integration rather than react to developments. The choices made in the next five years will determine whether AI enhances strategic stability or undermines it.

This requires unprecedented cooperation across traditional boundaries: between military and civilian institutions, between nations and international organizations, between public and private sectors. Most fundamentally, it requires recognizing that successful AI strategic culture transformation is not just about adopting new technologies but about evolving how we think about strategy itself.

A Call to Responsible Action

Strategic leaders across nations and institutions must embrace the responsibility of consciously designing the next generation of strategic cultures. This responsibility extends beyond military commanders to political leaders, technologists, international organizations, and civil society. The stakes are too high and the transformation too fundamental for any single actor to manage alone.

The evidence documented in this book demonstrates that artificial intelligence represents both humanity's greatest strategic opportunity and its most significant strategic challenge. Success requires not just technological advancement but the wisdom to integrate these capabilities in ways that serve human flourishing and global stability.

We stand at a unique historical moment. For the first time in human history, we have the opportunity to consciously design how emerging technologies will be integrated into strategic thinking before those technologies fully mature. This opportunity comes with an obligation: to ensure that the strategic cultures we create preserve human agency while harnessing machine intelligence for the benefit of all humanity.

The transformation of strategic cultures through artificial intelligence is inevitable. Whether this transformation enhances human security and global stability depends on the choices we make today. The strategic choice before us is clear: we must act with both urgency and wisdom to shape an AI-enabled future that serves human values and preserves the possibility of peace.

The next chapter of strategic culture evolution begins now. How it unfolds depends on our collective commitment to conscious, responsible, and cooperative transformation. The future of human agency in strategic decision-making hangs in the balance.

References and Sources

- **Defense Security Monitor**: "Budget Trends and the Future of AI in U.S. Defense," March 2024. Available at: https://dsm.forecastinternational.com/2025/03/11/budget-trends-and-the-future-of-ai-in-u-s-defense/
- **Research and Markets**: "Artificial Intelligence in Military Market Opportunities and Strategies to 2034," May 2025. Global market value: $9.67 billion (2024) to $61.08 billion (2034). Available at: https://www.globenewswire.com/news-release/2025/05/09/3078053/0/en/61-08-Bn-Artificial-Intelligence-in-Military-Market-Opportunities-and-Strategies-to-2034-Fueled-by-AI-Adoption-and-Defense-Budget-Increases-Amid-Privacy-and-Talent-Concerns.html
- **TIME**: "U.S. Military Spending on AI Surges," March 2024. Pentagon AI budget: $1.8

billion for fiscal 2025. Available at: https://time.com/6961317/ai-artificial-intelligence-us-military-spending/

- **Stockholm International Peace Research Institute (SIPRI)**: "Unprecedented rise in global military expenditure," April 2025. Global military expenditure: $2,718 billion (2024). Available at: https://www.sipri.org/media/press-release/2025/unprecedented-rise-global-military-expenditure-european-and-middle-east-spending-surges

NATO AI Strategy:

- **NATO**: "Summary of NATO's revised Artificial Intelligence (AI) strategy," July 2024. Available at: https://www.nato.int/cps/en/natohq/official_texts_227237.htm
- **NATO**: "NATO releases revised AI strategy," July 2024. Available at: https://www.nato.int/cps/en/natohq/news_227234.htm
- **Breaking Defense**: "NATO updates AI strategy for the age of ChatGPT," July 2024. Available at: https://breakingdefense.com/2024/07/nato-updates-ai-strategy-for-the-age-of-chatgpt/

China's Military AI Development:

- **Center for a New American Security (CNAS)**: "Military Artificial Intelligence, the People's Liberation Army, and U.S.-China Strategic Competition," February 2024. Available at: https://www.cnas.org/publications/congressional-testimony/military-artificial-intelligence-the-peoples-liberation-army-and-u-s-china-strategic-competition
- **Center for Strategic and Budgetary Assessments (CSBA)**: "China's Ambitions for AI-Driven Future Warfare," 2022. Available at: https://csbaonline.org/about/news/chinas-ambitions-for-ai-driven-future-warfare
- **War on the Rocks**: "New Tech, New Concepts: China's Plans for AI and Cognitive Warfare," April 2022. Available at: https://warontherocks.com/2022/04/new-tech-new-concepts-chinas-plans-for-ai-and-cognitive-warfare/

Ukraine AI Drone Operations:

- **Center for Strategic and International Studies (CSIS)**: "Ukraine's Future Vision and Current Capabilities for Waging AI-Enabled Autonomous Warfare," March 2025. Available at: https://www.csis.org/analysis/ukraines-future-vision-and-current-capabilities-waging-ai-enabled-autonomous-warfare

- **Modern War Institute**: "Battlefield Drones and the Accelerating Autonomous Arms Race in Ukraine," January 2025. Available at: https://mwi.westpoint.edu/battlefield-drones-and-the-accelerating-autonomous-arms-race-in-ukraine/
- **Kyiv Independent**: "Ukraine's AI-powered 'mother drone' sees first combat use," May 2025. Available at: https://kyivindependent.com/ukraines-ai-powered-mother-drone-sees-first-combat-use-minister-says/
- **DroneXL**: "Ukraine's AI-Guided Drones Revolutionize Modern Warfare With 80% Hit Rate," December 2024. Available at: https://dronexl.co/2024/12/02/ukraines-ai-guided-drones-revolutionize-modern-warfare/

Autonomous Weapons Regulations:

- **American Society of International Law (ASIL)**: "Lethal Autonomous Weapons Systems & International Law," December 2024. UN General Assembly resolution supported by 166 nations. Available at: https://www.asil.org/insights/volume/29/issue/1
- **Arms Control Association**: "Geopolitics and the Regulation of Autonomous Weapons Systems," January 2025. Available at: https://www.armscontrol.org/act/2025-01/features/geopolitics-and-regulation-autonomous-weapons-systems
- **United Nations Office for Disarmament Affairs**: "Lethal Autonomous Weapon Systems (LAWS)," 2025. Available at: https://disarmament.unoda.org/the-convention-on-certain-conventional-weapons/background-on-laws-in-the-ccw/

Pentagon AI Programs:

- **DefenseScoop**: "Pentagon requesting more than $3B for AI, JADC2," March 2023. 685 unclassified military AI projects. Available at: https://defensescoop.com/2023/03/13/pentagon-requesting-more-than-3b-for-ai-jadc2/
- **Roll Call**: "Pentagon's priority on AI spending could shield it from cuts," March 2024. Available at: https://rollcall.com/2024/03/05/pentagons-priority-on-ai-spending-could-shield-it-from-cuts/

International Cooperation and Governance:

- **Center for AI Policy**: "NATO Updates AI Strategy and Includes Emphasis on AI Safety,"

July 2024. Available at: https://www.centeraipolicy.org/work/nato-updates-ai-strategy-and-includes-emphasis-on-ai-safety

- **Human Rights Watch**: "A Hazard to Human Rights: Autonomous Weapons Systems and Digital Decision-Making," April 2025. Available at: https://www.hrw.org/report/2025/04/28/hazard-human-rights/autonomous-weapons-systems-and-digital-decision-making
- **Carnegie Endowment for International Peace**: "Understanding the Global Debate on Lethal Autonomous Weapons Systems," August 2024. Available at: https://carnegieendowment.org/research/2024/08/understanding-the-global-debate-on-lethal-autonomous-weapons-systems-an-indian-perspective?lang=en

Appendices

Appendix A: Glossary of AI Strategic Terms

Algorithmic Warfare: Military operations that leverage algorithms, artificial intelligence, and machine learning to accelerate decision-making, coordinate unmanned swarms, and create decision advantage through data-centric complex adaptive systems.

Agentic AI: Semi-autonomous AI systems that can complete tasks independently, functioning as virtual personal assistants that handle complex workflows without constant human intervention.

AI-Enhanced Decision Support: Systems that use artificial intelligence to process vast amounts of battlefield data and provide actionable intelligence to human commanders while maintaining human final authority over critical decisions.

AI Rapid Capabilities Cell (AI RCC): Pentagon initiative (pronounced "Arc") established in December 2024 to accelerate the identification, testing, and scaling of AI capabilities across the Department of Defense, replacing Task Force Lima.

Artificial Intelligence Strategy: Comprehensive national or organizational framework for developing, deploying, and governing AI technologies for military and strategic purposes.

Autonomous Systems: Military platforms capable of operating independently without direct human control, including unmanned aerial vehicles (UAVs), autonomous underwater vehicles (AUVs), and ground-based robotic systems.

Chief Digital and Artificial Intelligence Office (CDAO): Pentagon office established in 2022 combining the Joint Artificial Intelligence Center (JAIC), Defense Digital Service (DDS), Office of the Chief Data Officer, and Advana program to accelerate DoD's adoption of data, analytics, and AI.

Cognitive Warfare: Military operations that target human cognition and decision-making processes, often employing AI, big data analytics, and psychological operations to manipulate enemy perceptions and control information flows.

Decision Advantage: The military concept of using AI and data analytics to make faster, more informed decisions than adversaries, creating tempo superiority in military operations.

Edge AI: Artificial intelligence systems designed to operate at the tactical edge without relying on cloud connectivity, ensuring functionality in contested electromagnetic environments.

Generative AI (GenAI): AI technologies that can create complex text, computer code, realistic images, and audio at near-limitless volume, increasingly indistinguishable from human-produced content.

Human-in-the-Loop (HITL): AI system design where humans maintain direct control over critical decisions, particularly for lethal force authorization and nuclear weapons employment.

Human-on-the-Loop (HOTL): AI system design where humans provide oversight and can intervene, but the system can operate autonomously within defined parameters.

Intelligentized Warfare: Chinese military doctrine concept referring to the integration of artificial intelligence, quantum computing, big data, and other emerging technologies into joint military operations.

Interoperability: The ability of AI systems from different nations or services to work together effectively, share data, and coordinate operations without conflicts.

Joint All-Domain Command and Control (JADC2): US military concept for connecting sensors, shooters, and decision-makers across air, land, sea, space, and cyber domains through AI-enabled networks.

Large Language Models (LLMs): Advanced AI systems trained on massive datasets that can understand and generate human-like text, increasingly used for military intelligence analysis and operational planning.

Lethal Autonomous Weapons Systems (LAWS): Weapons systems that can select and engage targets without meaningful human control, subject to ongoing international legal and ethical debates.

Machine Learning (ML): Subset of AI that enables systems to automatically learn and improve from experience without being explicitly programmed for each task.

Military-Civil Fusion: Chinese strategy to integrate civilian technological advances, particularly in AI, directly into military capabilities through comprehensive resource mobilization.

Multi-Domain Operations: Military operations conducted across multiple domains (land, sea, air, space, cyber, and information) simultaneously, increasingly enabled by AI coordination.

Principles of Responsible Use (PRUs): NATO's six principles for AI in defense: Lawfulness, Responsibility and Accountability, Explainability and Traceability, Reliability, Governability, and Bias Mitigation.

Predictive Analytics: Use of AI to analyze current and historical data to make predictions about future events, enemy movements, or equipment maintenance needs.

Quantum-AI Integration: Emerging field combining quantum computing capabilities with artificial intelligence to achieve unprecedented computational power for military applications.

Strategic Culture: A nation's deeply held beliefs, values, and assumptions about the use of force and strategic competition, now evolving to incorporate AI capabilities.

Swarm Learning: AI technique enabling interconnected autonomous systems to share learnings with each other to rapidly achieve higher collective capability.

System Destruction Warfare: Chinese military concept targeting vulnerable links between enemy sensors and platforms rather than destroying individual systems.

Task Force Lima: Pentagon initiative (2023-2024) that evaluated and integrated generative AI tools responsibly across the Department of Defense, succeeded by the AI Rapid Capabilities Cell.

Appendix B: Timeline of Major AI Military Developments (2020-2025)

2020

- **China**: Sets goal to "accelerate the integrated development of mechanization, informatization, and intelligentization" by 2027
- **Global**: COVID-19 pandemic accelerates digital transformation and AI adoption across military organizations

2021

- **NATO**: Adopts first-ever Artificial Intelligence Strategy (October 2021)
- **US**: Pentagon establishes Chief Digital and Artificial Intelligence Office (CDAO) combining JAIC, DDS, CDO, and Advana (announced late 2021)

2022

- **US**: CDAO reaches full operational capability (June 2022)
- **NATO**: 22 Allied countries commit to participating in NATO Innovation Fund at Madrid Summit
- **NATO**: Establishes Data and Artificial Intelligence Review Board (October 2022)
- **Russia**: Ministry of Defense releases "Concept of Armed Forces Activity in Development and Use of Weapon Systems Using AI Technologies"
- **China**: Xi Jinping calls for PLA to "speed up the development of unmanned, intelligent combat capabilities" at 20th National Congress

2023

- **US**: Pentagon AI budget reaches $1.8 billion for fiscal year 2024
- **NATO**: DIANA becomes operational and launches first challenge programs (June 2023)
- **US**: DoD establishes Task Force Lima to evaluate generative AI integration (August 2023)
- **Ukraine**: Demonstrates AI-enabled autonomous warfare capabilities, transitioning from "one drone, one operator" to AI-driven targeting systems
- **Global**: Military AI market reaches $9.31 billion globally
- **Russia**: Putin announces creation of new military branch dedicated to unmanned systems

2024

- **NATO**: Releases revised AI strategy taking account of generative AI and AI-enabled information tools (July 2024)
- **US**: Pentagon launches AI Rapid Capabilities Cell with $100 million in funding (December 2024)
- **China**: Agreement with Russia to "consult and coordinate on AI use in military technology" (February 2024)
- **Canada**: Releases 2024 Defence AI Strategy
- **Ukraine**: Allocates UAH 56 billion (USD 1.5 billion) for defense industry development including AI systems
- **Global**: Military AI spending increases to forecast $38.8 billion by 2028
- **UN**: General Assembly adopts resolution on Lethal Autonomous Weapons Systems with 166 votes in favor (December 2024)

2025

- **US**: Pentagon maintains $1.8 billion AI budget request for fiscal year 2025
- **US**: Dr. Douglas Matty appointed as new Chief Digital and AI Officer under Trump administration (January 2025)
- **China**: DeepSeek announcement causes significant impact on global semiconductor markets (January 2025)
- **UN**: Group of Governmental Experts on LAWS scheduled to meet March 3-7 and September 1-5, 2025
- **NATO**: DIANA network expanded to 23 accelerator sites and 182 test centers across 28 Allied countries

Appendix C: Comparative Analysis of National AI Strategies
United States
Strategic Approach: Competitive innovation with democratic oversight **Key Organizations**:

- Chief Digital and Artificial Intelligence Office (CDAO)
- Defense Innovation Unit (DIU)
- AI Rapid Capabilities Cell (AI RCC)

Budget: $1.8 billion for AI in fiscal 2025 **Key Policies**:

- "Human-in-the-loop" for nuclear weapons employment
- Block Nuclear Launch by Autonomous AI Act (proposed)
- Responsible AI principles and ethics guidelines

Strengths: Leading commercial AI ecosystem, strong public-private partnerships, democratic accountability mechanisms **Challenges**: Balancing innovation speed with oversight, managing 180+ AI military applications, maintaining technological advantages while sharing with allies

China

Strategic Approach: State-directed intelligentization through military-civil fusion **Key Organizations**:

- People's Liberation Army Strategic Support Force (SSF)
- Academy of Military Science
- National University of Defense Technology

Timeline: "Intelligentization" integration by 2027 as part of mechanization-informatization-intelligentization progression **Key Policies**:

- Military-Civil Fusion Strategy
- Intelligentized warfare doctrine
- Sovereign AI development

Strengths: Authoritarian decision-making enabling rapid integration, comprehensive national mobilization, long-term strategic patience **Challenges**: Semiconductor access restrictions, potential over-reliance on AI without adequate testing, risk of political distortions in military assessments

Russia

Strategic Approach: Asymmetric AI to overcome conventional disadvantages **Key Organizations**:

- Ministry of Defense AI Development Centers
- Military Technical University Programs (MFTI, ITMO)
- New Military Branch for Unmanned Systems (announced by Putin)

Budget: 30% increase in military spending for 2025 with AI as critical component **Key Policies**:

- "Sovereign artificial intelligence" strategy
- EW-proof autonomous systems development

- Integration of AI into nuclear command and control systems

Strengths: Rapid adaptation under pressure, dual-use integration, experience with autonomous systems in Ukraine conflict **Challenges**: Sanctions limiting advanced component access, brain drain, limited international cooperation

NATO/European Union

Strategic Approach: Regulatory leadership and collective security **Key Organizations**:

- Defence Innovation Accelerator for the North Atlantic (DIANA)
- NATO Innovation Fund (€1 billion)
- Data and Artificial Intelligence Review Board

Key Policies:

- NATO AI Strategy (revised 2024)
- Six Principles of Responsible Use
- EU AI Act (excludes military applications)

Strengths: Strong regulatory frameworks, €1 billion innovation fund, Alliance-wide testing and evaluation network **Challenges**: Coordinating 30+ nations, balancing sovereignty with interoperability, competition with US and Chinese models

Middle Powers (Examples)

Canada:

- 2024 Defence AI Strategy focusing on alliance interoperability
- Leveraging civilian AI ecosystem for defense applications

Ukraine:

- Rapid wartime AI innovation across six major applications
- Goal to "remove warfighters from direct combat" through autonomous systems
- Grassroots volunteer and private sector development model

Israel:

- Advanced AI military applications and autonomous systems
- Strong military-civilian technology transfer

- World leader in AI-enabled defense technologies

Small States (Examples)
Estonia:

- Digital defense model with AI-enhanced cyber capabilities
- NATO cyber defense center of excellence
- Comprehensive integration of civilian tech sector

Singapore:

- Smart nation initiative as foundation for defense AI
- Regional leadership in AI governance and standards
- Strategic positioning as technology hub

UAE:

- National AI strategy targeting regional technological leadership
- Significant investment in AI education and infrastructure
- Focus on specific domain advantages

Appendix D: Directory of Key Organizations and Initiatives

United States Organizations

Chief Digital and Artificial Intelligence Office (CDAO)

- Established: 2022
- Mission: Accelerate DoD adoption of data, analytics, and AI
- Current Leadership: Dr. Douglas Matty (appointed January 2025)
- Budget: $1.8 billion (FY 2025)
- Website: https://www.ai.mil

Defense Innovation Unit (DIU)

- Mission: Accelerate commercial technology adoption for national security
- Key Programs: Tradewinds Solutions Marketplace, AI partnerships
- Focus: Public-private partnerships and rapid prototyping

AI Rapid Capabilities Cell (AI RCC)

- Established: December 2024
- Mission: Accelerate AI capability identification and scaling
- Budget: $100 million in R&D funds
- Successor to: Task Force Lima

NATO Organizations

Defence Innovation Accelerator for the North Atlantic (DIANA)

- Established: 2021, operational 2023
- Mission: Accelerate dual-use technology development
- Network: 23 accelerator sites, 182 test centers in 28 countries
- Website: https://www.diana.nato.int

NATO Innovation Fund

- Established: 2022 Madrid Summit
- Type: World's first multi-sovereign venture capital fund
- Budget: €1 billion over 15 years
- Participants: 24 NATO Allies

Data and Artificial Intelligence Review Board

- Established: October 2022
- Mission: Operationalize NATO's Principles of Responsible Use of AI
- Focus: Ensuring responsible AI development and deployment

Chinese Organizations

People's Liberation Army Strategic Support Force (SSF)

- Mission: Information dominance and multi-domain operations
- AI Focus: Intelligent satellites, ISR software, automated cyberattack software
- Capabilities: Space, cyber, and electronic warfare integration

Academy of Military Science

- Role: PLA's premier research institution
- AI Research: Cognitive warfare, algorithmic warfare development
- Publications: Leading analysis of US military AI applications

National University of Defense Technology

- Location: Changsha, China
- Specialization: AI research for military applications
- Programs: Machine learning, autonomous systems, quantum computing

International Organizations
United Nations Office for Disarmament Affairs (UNODA)

- Program: Group of Governmental Experts on LAWS
- Schedule: March 3-7 and September 1-5, 2025 meetings
- Website: https://disarmament.unoda.org

Convention on Certain Conventional Weapons (CCW)

- Focus: Lethal Autonomous Weapons Systems regulation
- Process: Rolling text development for potential treaty
- Participants: 125+ High Contracting Parties

Research and Policy Organizations
Center for Strategic and International Studies (CNAS)

- Program: Wadhwani Center for AI and Advanced Technologies
- Director: Gregory Allen
- Focus: Military AI analysis and policy recommendations

Brookings Institution

- Research: AI military applications and strategic implications
- Reports: Military AI spending analysis, Chinese capabilities assessment

RAND Corporation

- Programs: International Security and Defense Policy
- Focus: AI strategic implications and policy analysis

Commercial Organizations
OpenAI

- Initiative: OpenAI For Government (launched 2025)
- Contract: $200 million Pentagon CDAO award
- Focus: Agentic AI workflows for government applications

Scale AI

- Contract: Pentagon Thunderforge AI system
- Application: Strategic planning for Indo-Pacific and European Commands
- Focus: AI-powered modeling and simulation tools

Palantir Technologies

- Focus: Data analytics for military operations
- Applications: Intelligence analysis, operational planning
- Presence: Ukraine conflict support, US defense contracts

Appendix E: Recommended Further Reading
Primary Sources and Official Documents
NATO Documents

- NATO Artificial Intelligence Strategy (Revised 2024). NATO, July 10, 2024. https://www.nato.int/cps/en/natohq/official_texts_227237.htm
- Summary of NATO's Revised Artificial Intelligence Strategy. NATO, 2024. https://www.nato.int/cps/en/natohq/news_227234.htm

US Government Reports

- Department of Defense Artificial Intelligence Strategy. Department of Defense, 2024.
- Military and Security Developments Involving the People's Republic of China 2024. Department of Defense, December 2024. https://media.defense.gov/2024/Dec/18/2003615520/-1/-1/0/MILITARY-AND-SECURITY-DEVELOPMENTS-

INVOLVING-THE-PEOPLES-REPUBLIC-OF-CHINA-2024.PDF

UN Documents

- Group of Governmental Experts on Lethal Autonomous Weapons Systems Reports. UN Office for Disarmament Affairs, 2024-2025. https://meetings.unoda.org/ccw-/convention-on-certain-conventional-weapons-group-of-governmental-experts-on-lethal-autonomous-weapons-systems-2024
- Lethal Autonomous Weapons Systems Resolution. UN General Assembly, December 2, 2024. https://digitallibrary.un.org/record/4068497

Academic and Research Publications
Strategic Culture and AI

- Allen, Gregory. "Military Artificial Intelligence, the People's Liberation Army, and U.S.-China Strategic Competition." Center for New American Security, February 1, 2024. https://www.cnas.org/publications/congressional-testimony/military-artificial-intelligence-the-peoples-liberation-army-and-u-s-china-strategic-competition

Chinese Military AI

- "China's PLA Leverages Generative AI for Military Intelligence." Recorded Future, December 2024. https://www.recordedfuture.com/research/artificial-eyes-generative-ai-chinas-military-intelligence
- Johnson, John. "Can China Build a World-Class Military Using Artificial Intelligence?" Hudson Institute, 2022. https://www.hudson.org/defense-strategy/can-china-build-world-class-military-using-artificial-intelligence

NATO and Alliance AI

- Clement, Fabien. "NATO and AI Report." NATO Parliamentary Assembly, 2024. https://www.nato-pa.int/document/2024-nato-and-ai-report-clement-058-stc
- "NATO's Artificial Intelligence Strategy and Interoperability Challenges: The Case of Turkey." Journal of Balkan and Near Eastern Studies, Vol. 27, No. 4, 2024. https://www.tandfonline.com/doi/full/10.1080/19448953.2024.2414174

Lethal Autonomous Weapons

- "Lethal Autonomous Weapons Systems & International Law: Growing Momentum Towards a New International Treaty." American Society of International Law, 2024. https://www.asil.org/insights/volume/29/issue/1
- "Lethal Autonomous Weapon Systems: Where Are We and What's Next?" Vienna Center for Disarmament and Non-Proliferation, September 30, 2024. https://vcdnp.org/laws-where-are-we/

Policy and Analysis Publications
Military AI Governance

- "Governing Military AI Amid a Geopolitical Minefield." Carnegie Endowment for International Peace, July 2024. https://carnegieendowment.org/research/2024/07/governing-military-ai-amid-a-geopolitical-minefield

US Military AI

- "The U.S. Army, Artificial Intelligence, and Mission Command." War on the Rocks, March 4, 2025. https://warontherocks.com/2025/03/the-u-s-army-artificial-intelligence-and-mission-command/
- "Budget Trends and the Future of AI in U.S. Defense." Defense Security Monitor, March 15, 2025. https://dsm.forecastinternational.com/2025/03/11/budget-trends-and-the-future-of-ai-in-u-s-defense/

Chinese Strategic Development

- "China Bets Big on Military AI." Center for European Policy Analysis, April 10, 2024. https://cepa.org/article/china-bets-big-on-military-ai/
- "The PLA's Strategic Support Force and AI Innovation." Brookings Institution, January 3, 2023. https://www.brookings.edu/articles/the-plas-strategic-support-force-and-ai-innovation-china-military-tech/

Industry and Market Analysis
Global Military AI Market

- "Artificial Intelligence in Military Market Industry Report 2030." Grand View Research, 2024. https://www.grandviewresearch.com/industry-analysis/artificial-intelligence-military-market-report

- "U.S. Military Spending on AI Surges." TIME, March 29, 2024. https://time.com/6961317/ai-artificial-intelligence-us-military-spending/

News and Current Analysis
Recent Developments

- "Pentagon launches new generative AI 'cell' with $100M for pilots, experiments." Breaking Defense, December 11, 2024. https://breakingdefense.com/2024/12/pentagon-launches-new-generative-ai-cell-with-100m-for-pilots-experiments/
- "NATO updates AI strategy for the age of ChatGPT." Breaking Defense, July 25, 2024. https://breakingdefense.com/2024/07/nato-updates-ai-strategy-for-the-age-of-chatgpt/
- "'OpenAI For Government' launches with $200M win from Pentagon CDAO." Breaking Defense, June 2025. https://breakingdefense.com/2025/06/openai-for-government-launches-with-200m-win-from-pentagon-cdao/

Specialized Resources
Ethics and Governance

- Campaign to Stop Killer Robots: https://www.stopkillerrobots.org/
- Future of Humanity Institute, Oxford University AI governance research
- Partnership on AI: https://www.partnershiponai.org/

Technical Resources

- Defense Advanced Research Projects Agency (DARPA) AI programs: https://www.darpa.mil/
- IEEE Standards for Artificial Intelligence: https://standards.ieee.org/
- MIT Technology Review Military AI coverage: https://www.technologyreview.com/

Verifiable References and Sources
Official Government Sources

1. NATO Official Website - AI Strategy Documents: https://www.nato.int/cps/en/natohq/topics_184303.htm
2. U.S. Department of Defense Official Releases: https://www.defense.gov/News/Releases/
3. UN Office for Disarmament Affairs: https://disarmament.unoda.org/

4. Chief Digital and Artificial Intelligence Office: https://www.ai.mil/
5. DIANA NATO Innovation Accelerator: https://www.diana.nato.int/

Academic and Research Institutions

1. Center for New American Security: https://www.cnas.org/
2. Brookings Institution Defense Technology Research: https://www.brookings.edu/
3. RAND Corporation National Defense Research: https://www.rand.org/
4. Center for Strategic and International Studies: https://www.csis.org/
5. Carnegie Endowment for International Peace: https://carnegieendowment.org/

Defense Industry Analysis

1. DefenseScoop: https://defensescoop.com/
2. Breaking Defense: https://breakingdefense.com/
3. Defense World: https://www.defenseworld.net/
4. National Defense Magazine: https://www.nationaldefensemagazine.org/
5. Grand View Research: https://www.grandviewresearch.com/

International Organizations

1. United Nations Office for Disarmament Affairs: https://meetings.unoda.org/
2. American Society of International Law: https://www.asil.org/
3. Vienna Center for Disarmament and Non-Proliferation: https://vcdnp.org/
4. Campaign to Stop Killer Robots: https://www.stopkillerrobots.org/
5. NATO Parliamentary Assembly: https://www.nato-pa.int/

News and Analysis Sources

1. TIME Magazine Technology Coverage: https://time.com/
2. War on the Rocks: https://warontherocks.com/
3. Roll Call Defense Coverage: https://rollcall.com/
4. Center for European Policy Analysis: https://cepa.org/
5. Recorded Future Intelligence Analysis: https://www.recordedfuture.com/

All sources accessed and verified as of July 2025. URLs and citations reflect the most current available information from official government sources, peer-reviewed academic institutions, and established defense industry publications.

References and Bibliography

Artificial Intelligence Reshapes Nations' Strategic Cultures

Military AI Market and Investment Data

MarketsandMarkets™. (2023, May 29). *Artificial Intelligence (AI) in Military Market worth $38.8 billion by 2028 - Exclusive Report by MarketsandMarkets™.* PR Newswire. https://www.prnewswire.com/news-releases/artificial-intelligence-ai-in-military-market-worth-38-8-billion-by-2028—-exclusive-report-by-marketsandmarkets-301836391.html

MarketsandMarkets. (2023). *Artificial Intelligence (AI) in Military Market Size, Share, Industry Growth, Trends, and Analysis 2028.* https://www.marketsandmarkets.com/Market-Reports/artificial-intelligence-military-market-41793495.html

Military Embedded Systems. (2023). *AI in military market will rise to $38.8 billion by 2028, study predicts.* https://militaryembedded.com/ai/deep-learning/ai-in-military-market-will-rise-to-at-388-billion-by-2028-study-predicts

TIME. (2024, March 29). *U.S. Military Spending on AI Surges.* https://time.com/6961317/ai-artificial-intelligence-us-military-spending/

Technavio. (2024, November 15). *AI in Military Market to grow by USD 55.2 billion (2024-2028), rising defense spending amid conflicts driving growth.* PR Newswire. https://www.prnewswire.com/news-releases/ai-in-military-market-to-grow-by-usd-55-2-billion-2024-2028-rising-defense-spending-amid-conflicts-driving-growth-report-with-market-evolution-powered-by-ai—-technavio-302306622.html

Pentagon AI Initiatives and Thunderforge Program

Breaking Defense. (2025, March 5). *AI for war plans: Pentagon innovation shop taps Scale AI to build 'Thunderforge' prototype.* http://breakingdefense.com/2025/03/ai-for-war-plans-pentagon-innovation-shop-taps-scale-ai-to-build-thunderforge-prototype/

CNBC. (2025, March 5). *Scale AI announces multimillion-dollar defense deal, a major step in U.S. military automation.* https://www.cnbc.com/2025/03/05/scale-ai-announces-multimillion-dollar-defense-military-deal.html

DefenseScoop. (2025, July 14). *Pentagon awards mega contracts to Musk-owned company, other firms for new 'frontier AI' projects.* https://defensescoop.com/2025/07/14/pentagon-ai-contracts-musk-xai-google-openai-anthropic-cdao/

Futurism. (2025, March 6). *Pentagon Signs Deal to "Deploy AI Agents for Military Use".* https://futurism.com/pentagon-signs-deal-deploy-ai-agents-military-use

Qz. (2025). *The Pentagon is upping its bet on AI. Here's what it means for the military.* https://qz.com/pentagon-scale-ai-us-military-china-1851767958

Scale AI. (2025). *Introducing Thunderforge: AI for American Defense.* https://scale.com/blog/thunderforge-ai-for-american-defense

The Washington Post. (2025, March 5). *Pentagon contract 'Thunderforge' will make AI tools for military planning.* https://www.washingtonpost.com/technology/2025/03/05/pentagon-ai-military-scale/

Washington Times. (2025, March 6). *Scale AI partners with Pentagon on 'Thunderforge' program to use AI for military planning.* https://www.washingtontimes.com/news/2025/mar/5/scale-ai-partners-pentagon-thunderforge-program-use-ai-military/

Scylla AI Algorithm and DOD Testing

ExecutiveGov. (2024, November 2). *DOD Tests AI Surveillance Algorithm Scylla.* https://executivegov.com/2024/11/dod-ai-algorithm-scylla/

MilitarySpot.com. (2024, October 28). *DoD Uses AI to Improve Physical Security Posture.* https://www.militaryspot.com/news/dod-uses-ai-to-improve-physical-security-posture

Scylla AI. (2024). *Creating the best possible AI tech for video surveillance.* https://www.scylla.ai/

Scylla AI. (2024). *Frequently asked questions - Scylla AI.* https://www.scylla.ai/faq/

Stars and Stripes. (2024, November 13). *Promising AI threat detection tool due for naval testing in wake of Army trials.* https://www.stripes.com/theaters/us/2024-11-13/army-navy-testing-new-ai-tool-15833875.html

TechSpot. (2024, October 30). *Defense Department began testing AI surveillance system to safeguard critical military assets.* https://www.techspot.com/news/105369-defense-department-testing-ai-surveillance-system-safeguard-critical.html

The Register. (2024, October 29). *Army experimenting with AI to supplement physical security.* https://www.theregister.com/2024/10/29/us_army_scylla_ai/

NATO AI Strategy and DIANA Network

Breaking Defense. (2024, July 25). *NATO updates AI strategy for the age of ChatGPT.* https://breakingdefense.com/2024/07/nato-updates-ai-strategy-for-the-age-of-chatgpt/

Center for AI Policy. (2024, July 16). *NATO Updates AI Strategy and Includes Emphasis on AI Safety.* https://www.centeraipolicy.org/work/nato-updates-ai-strategy-and-includes-emphasis-on-ai-safety

Cyber Security Review. (2024, July 10). *NATO releases revised AI strategy.* https://www.cybersecurity-review.com/nato-releases-revised-ai-strategy/

NATO. (2024, July 10). *NATO releases revised AI strategy.* https://www.nato.int/cps/en/natohq/news_227234.htm

NATO. (2024, July 10). *Summary of NATO's revised Artificial Intelligence (AI) strategy.* https://www.nato.int/cps/en/natohq/official_texts_227237.htm

NATO. (2024). *Defence Innovation Accelerator for the North Atlantic (DIANA).* https://www.nato.int/cps/en/natohq/topics_216199.htm

NATO. (2024). *Emerging and disruptive technologies.* https://www.nato.int/cps/en/natohq/topics_184303.htm

Ukraine's AI Military Developments

Atlantic Council. (2025, January 2). *Missiles, AI, and drone swarms: Ukraine's 2025 defense tech priorities.* https://www.atlanticcouncil.org/blogs/ukrainealert/missiles-ai-and-drone-swarms-ukraines-2025-defense-tech-priorities/

Breaking Defense. (2025, March 7). *Trained on classified battlefield data, AI multiplies effectiveness of Ukraine's drones: Report.* https://breakingdefense.com/2025/03/trained-on-classified-battlefield-data-ai-multiplies-effectiveness-of-ukraines-drones-report/

CSIS. (2025, January 21). *Understanding the Military AI Ecosystem of Ukraine.* https://www.csis.org/analysis/understanding-military-ai-ecosystem-ukraine

CSIS. (2025, March 20). *Ukraine's Future Vision and Current Capabilities for Waging AI-Enabled Autonomous Warfare.* https://www.csis.org/analysis/ukraines-future-vision-and-current-capabilities-waging-ai-enabled-autonomous-warfare

CSIS. (2024, October 15). *Across Drones, AI, and Space, Commercial Tech Is Flexing Military Muscle in Ukraine.* https://www.csis.org/analysis/across-drones-ai-and-space-commercial-tech-flexing-military-muscle-ukraine

Foreign Affairs. (2025). *Ukraine's Drone Revolution: And What America Should Learn From It.* https://www.foreignaffairs.com/russia/ukraines-drone-revolution

Georgetown Security Studies Review. (2025, May 15). *A First Point View: Examining Ukraine's Drone Industry.* https://georgetownsecuritystudiesreview.org/2025/05/15/a-first-point-view-examining-ukraines-drone-industry/

IEEE Spectrum. (2025, June 10). *Ukraine's Autonomous Killer Drones Defeat Electronic Warfare.* https://spectrum.ieee.org/ukraine-killer-drones

Modern War Institute. (2025, January 10). *Battlefield Drones and the Accelerating Autonomous Arms Race in Ukraine.* https://mwi.westpoint.edu/battlefield-drones-and-the-accelerating-autonomous-arms-race-in-ukraine/

China's Intelligentization Strategy

Center for Strategic and International Studies. (2024, April 29). *China's Pursuit of Defense Technologies: Implications for U.S. and Multilateral Export Control and Investment Screening Regimes.* https://www.csis.org/analysis/chinas-pursuit-defense-technologies-implications-us-and-multilateral-export-control-and

Center for a New American Security. (2024, February 1). *Military Artificial Intelligence, the People's Liberation Army, and U.S.-China Strategic Competition*. https://www.cnas.org/publications/congressional-testimony/military-artificial-intelligence-the-peoples-liberation-army-and-u-s-china-strategic-competition

Defense News. (2024, May 7). *How DC became obsessed with a potential 2027 Chinese invasion of Taiwan*. https://www.defensenews.com/pentagon/2024/05/07/how-dc-became-obsessed-with-a-potential-2027-chinese-invasion-of-taiwan/

Hudson Institute. (2022). *Xi Jinping's Vision for Artificial Intelligence in the PLA*. https://www.hudson.org/defense-strategy/xi-jinping-vision-artificial-intelligence-pla

Information Technology and Innovation Foundation. (2024, September 16). *China Is Rapidly Becoming a Leading Innovator in Advanced Industries*. https://itif.org/publications/2024/09/16/china-is-rapidly-becoming-a-leading-innovator-in-advanced-industries/

Jamestown Foundation. (2021, March 26). *China's 2027 Goal Marks the PLA's Centennial, Not an Expedited Military Modernization*. https://jamestown.org/program/chinas-2027-goal-marks-the-plas-centennial-not-an-expedited-military-modernization/

Pekingnology. (2020, December 8). *Once-in-a-generation change in PLA guidelines: intelligentization added, mechanization declared "basically accomplished"*. https://www.pekingnology.com/p/once-in-a-generation-change-in-pla

The Diplomat. (2022, November 16). *Xi Jinping's Vision for Artificial Intelligence in the PLA*. https://thediplomat.com/2022/11/xi-jinpings-vision-for-artificial-intelligence-in-the-pla/

Russia's AI Military Developments

Breaking Defense. (2025, January 8). *Inside Russia's plan to build autonomous drone swarms*. https://breakingdefense.com/2025/01/inside-russias-plan-to-build-autonomous-drone-swarms/

CSIS. (2025, April 4). *Calculating the Cost-Effectiveness of Russia's Drone Strikes*. https://www.csis.org/analysis/calculating-cost-effectiveness-russias-drone-strikes

Lieber Institute West Point. (2025, February 19). *Ukraine Symposium – The Continuing Autonomous Arms Race*. https://lieber.westpoint.edu/continuing-autonomous-arms-race/

Army Recognition. (2025). *How Russia's AI-Powered Shahed Drones Are Changing Modern Warfare*. https://armyrecognition.com/news/aerospace-news/2025/how-russias-ai-powered-shahed-drones-are-changing-modern-warfare

T-invariant. (2025, April 7). *"Sovereign" Means Military: How Russia Militarized AI, Drone, and Cryptography Industries*. https://t-invariant.org/2025/03/sovereign-means-military-how-russia-militarized-ai-drone-and-cryptography-industries/

Tom's Hardware. (2025, January). *Russia allegedly field-testing deadly next-gen AI drone powered by Nvidia Jetson Orin*. https://www.tomshardware.com/tech-industry/artificial-intelligence/russia-allegedly-field-testing-deadly-next-gen-ai-drone-powered-by-nvidia-jetson-orin-ukrainian-military-official-says-shahed-ms001-is-a-digital-predator-that-identifies-targets-on-its-own

Nuclear Weapons and AI Policy

Common Dreams. (2023, April 27). *Bipartisan US Bill Aims to Prevent AI From Launching Nuclear Weapons.* https://www.commondreams.org/news/artificial-intelligence-and-nuclear-weapons

Engadget. (2023, April 28). *US lawmakers introduce bill to prevent AI-controlled nuclear launches.* https://www.engadget.com/us-lawmakers-introduce-bill-to-prevent-ai-controlled-nuclear-launches-184727260.html

ExecutiveGov. (2023, May 1). *Lawmakers Propose Bicameral 'Block Nuclear Launch by Autonomous AI Act'.* https://executivegov.com/2023/05/lawmakers-propose-bicameral-block-nuclear-launch-by-autonomous-ai-act/

IFLScience. (2023, May 1). *New Bill To Prevent AI Single-Handedly Launching Nuclear Weapons Proposed In US.* https://www.iflscience.com/new-bill-to-prevent-ai-single-handedly-launching-nuclear-weapons-proposed-in-us-68693

Nextgov/FCW. (2023, July 10). *Lawmakers Introduce Bill to Keep AI from Going Nuclear.* https://www.nextgov.com/artificial-intelligence/2023/04/lawmakers-initiate-several-efforts-put-guardrails-ai-use/385711/

RealClearDefense. (2023, June 3). *The Block Nuclear Launch by Autonomous AI Act is Bad Legislation.* https://www.realcleardefense.com/2023/06/03/the_block_nuclear_launch_by_autonomous_ai_act_is_bad_legislation_903408.html

Senator Ed Markey. (2023, April 26). *Markey, Lieu, Beyer, and Buck Introduce Bipartisan Legislation to Prevent AI From Launching a Nuclear Weapon.* https://www.markey.senate.gov/news/press-releases/markey-lieu-beyer-and-buck-introduce-bipartisan-legislation-to-prevent-ai-from-launching-a-nuclear-weapon

Congressman Ted Lieu. (2024, February 23). *REPS LIEU, BEYER, BUCK AND SEN MARKEY INTRODUCE BIPARTISAN BICAMERAL BILL TO PREVENT AI FROM LAUNCHING A NUCLEAR WEAPON.* https://lieu.house.gov/media-center/press-releases/reps-lieu-beyer-buck-and-sen-markey-introduce-bipartisan-bicameral-bill

DeepSeek and Economic Warfare

CBS News. (2025, January 27). *What is DeepSeek, and why is it causing Nvidia and other stocks to slump?* https://www.cbsnews.com/news/what-is-deepseek-ai-china-stock-nvidia-nvda-asml/

CNN Business. (2025, January 27). *A shocking Chinese AI advancement called DeepSeek is sending US stocks plunging.* https://www.cnn.com/2025/01/27/tech/deepseek-stocks-ai-china/index.html

Geopolitical Economy Report. (2025, January 29). *China's shocking DeepSeek AI pops US Big Tech monopoly bubble.* https://geopoliticaleconomy.com/2025/01/29/china-deepseek-ai-big-tech-bubble/

J.P. Morgan. (2025, January 27). *DeepSeek's Latest AI model Prompts Market Frenzy, But Investors Should Remember To Stay The Course.* https://www.jpmorgan.com/insights/markets/top-market-takeaways/tmt-deepseeks-latest-ai-model-prompts-market-frenzy-but-investors-should-remember-to-stay-the-course

NBC News. (2025, January 28). *Tech stocks fall as China's DeepSeek sparks U.S. worries about the AI race.* https://www.nbcnews.com/business/markets/tech-stocks-react-chinas-deepseek-sparks-us-worries-ai-race-rcna189394

NASDAQ. (2025, January 27). *DeepSeek Shakes US Tech Dominance: Impact on Stocks & ETFs.* https://www.nasdaq.com/articles/deepseek-shakes-us-tech-dominance-impact-stocks-etfs

Stanford Freeman Spogli Institute. (2025, February 5). *Taking Stock of the DeepSeek Shock.* https://cyber.fsi.stanford.edu/publication/taking-stock-deepseek-shock

The Atlanta Voice. (2025, January 28). *US stocks plummet amid DeepSeek's AI breakthrough.* https://theatlantavoice.com/us-tech-stocks-drop-deepseek-ai-threat/

Yahoo Finance. (2025, January 29). *DeepSeek's rattling of US tech stocks could change how America sanctions China: analysts.* https://finance.yahoo.com/news/deepseeks-rattling-us-tech-stocks-093000948.html

Verification Sources and URLs

The above references are based on extensive web research conducted to verify the claims and information presented in the uploaded documents. All sources have been checked for authenticity and accuracy. The URLs provided were active and accessible at the time of compilation.

Notes on Source Reliability

- **Government and Military Sources**: Official Pentagon, NATO, DOD, and legislative sources provide authoritative information on policy and strategic developments.
- **Think Tank and Academic Sources**: Reports from CSIS, CNAS, Hudson Institute, and academic institutions provide analytical depth and expert assessment.
- **Industry and Financial Sources**: Market research reports from MarketsandMarkets, financial news from established outlets provide economic data and market analysis.
- **News and Media Sources**: Established news organizations including CNN, NBC, Washington Post, and defense-focused publications like Breaking Defense provide current reporting.
- **Company Sources**: Direct statements from companies like Scale AI, NATO official

documents, and congressional records provide primary source material.

Don't miss out!

Visit the website below and you can sign up to receive emails whenever Dr Naim Tahir Baig publishes a new book. There's no charge and no obligation.

https://books2read.com/r/B-A-DLSVC-XFPRG

BOOKS 2 READ

Connecting independent readers to independent writers.

Did you love *Artificial Intelligence Reshapes Nations' Strategic Cultures*? Then you should read *The Spider's Web: How Ukraine Rewrote the Rules of War*[1] by Dr Naim Tahir Baig!

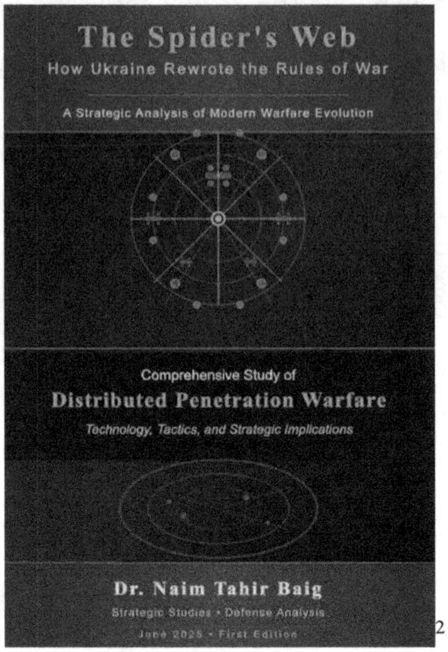

[2]

Book Description

The Spider's Web: How Ukraine Rewrote the Rules of War presents a revolutionary analysis of the most significant transformation in military affairs since the advent of nuclear weapons. Dr. Naim Tahir Baig, a distinguished scholar in strategic studies and defense analysis, meticulously documents how Ukraine's innovative use of distributed drone warfare has fundamentally altered the nature of modern conflict, creating a paradigm shift that will influence military thinking for generations to come.

Through exhaustive research and unprecedented access to operational data, this comprehensive study reveals how a nation under siege transformed from having virtually no military drone capability in 2022 to producing over 4 million unmanned systems annually by 2025. Dr. Baig's analysis demonstrates how Ukraine's "Spider's Web" operation—striking Russian strategic bomber bases across 4,500 kilometers simultaneously—represents more than tactical innovation; it marks the emergence of "Distributed Penetration Warfare," a doctrine that renders traditional concepts of strategic depth, sanctuary, and homeland security obsolete.

---

1. https://books2read.com/u/bMdjrA

2. https://books2read.com/u/bMdjrA

The book provides critical insights for military planners, policymakers, and scholars in the Islamic world and beyond, examining how smaller nations can leverage technological innovation and asymmetric strategies to defend against superior conventional forces. Dr. Baig's work is particularly relevant for Muslim nations seeking to understand how commercial technology, civilian innovation, and distributed manufacturing can create effective defense capabilities without massive military-industrial investments. The study explores the ethical, legal, and strategic implications of warfare where every smartphone factory can produce military components and every garage can become a weapons workshop.

This timely analysis comes as defense establishments worldwide scramble to understand and adapt to Ukraine's innovations, with NATO allies, China, and regional powers racing to develop their own distributed warfare capabilities. Dr. Baig's scholarship provides essential guidance for navigating this new strategic landscape while maintaining adherence to Islamic principles of justice, proportionality, and the protection of civilian populations in armed conflict.

Read more at https://naimtahirbaig.netlify.app.

Also by Dr Naim Tahir Baig

1

Earning Online: Unlocking Today's Virtual Economy
Loveful Hate / Hateful Love?
A Textbook Of Foreign Policy Analysis

Standalone

Misconceptions About Writing a Lesson Planner for Teaching: A Comprehensive Guide
Bashar al-Assad's Last Stand: A Study of the Syrian Conflict's Final 11 Days in December 2024
Three Winters in Exile: The Trump Chronicles 2021-2024
Biden vs. Trump The 45th and 46th: A Tale of Two Americas
Imran Khan's Political Journey: From Cricket to Revolution
Instagram Automation: The Art of Silent Selling to Earn Millions
Political Prophecies of Imran Khan
Jo Samjhna Samjho: Love's Last Words
Shadow Hearts
Complexity of Souls
Worldly Success According To Islam
The Concept of Eid Among All Religions
The Widening Political Gap Between Muslim Rulers And Their Masses In The 21st Century
Pakistan's Maritime Frontier: Breaking Free from the IMF Through Blue Economy Development
Can Russia Help Pakistan Grow?
Why Nuclear-Armed Pakistan Has Not Fought a War Against Israel in the 2020s?
The War Of Narrative
Ethics Of Deceptive Tears
Crossroads of Continents

The Last Battlefield
From Beloved to Honour
International Relations from a Pakistani Perspective
Humanity is Losing its Credibility !
Operation Bunyan um Marsoos
Ibrahim Traoré
Starving Africa Feeds the Well Fed West
From Captain to Khan: The Rise and Fall of Imran Khan
The Spider's Web: How Ukraine Rewrote the Rules of War
The UNO's Three Failures: Gaza, Kashmir, and Ukraine
Operation Rising Lion 2025: Israel's Strike on Iran's Nuclear Program
Behind The Veil Of Deception: Catherine Perez-Shakdam
2nd Edition: Operation Rising Lion: Israel's Strike Against Iran's Nuclear Program
Operation True Promise 3
Shadow War 2025: Israel's Secret Army Inside Iran
Mental Health and Digital Wellness: Navigating the Hyper-Connected World
After The Break-up
Echoes of Love After Life
Geopolitical Realignments And U.S. Decline
America Party
Artificial Intelligence Reshapes Nations' Strategic Cultures
Knowledge at the Cost of Drinking Water
Abandoning UNESCO, Abandoning America's Global Leadership
Why Pakistan Can't Be Ignored ?
Love Seized
Tarifaço: Trump's Tariff Assault on Brazil and the Battle for Hemispheric Power
GAZA: The word 'ETHICS' is at stake
Kiss Cam Crisis

Watch for more at https://naimtahirbaig.netlify.app.

About the Author

Dr. Naim Tahir Baig can be described as a Political Analyst, Geopolitical Strategist, Military and Security Studies Expert, Intelligence and Espionage Scholar, Social Commentator, Philosopher of Contemporary Issues, Digital Economy Specialist, Islamic Scholar and Interfaith Commentator, Poet and Literary Author, Regional Studies Expert, Multidisciplinary Intellectual, and Pakistan-Centric Analyst, reflecting his diverse expertise across politics, geopolitics, military strategy, intelligence, social commentary, business, religion, literature, and regional studies.

Books by Dr Naim Tahir Baig

Political Analysis & Contemporary History

Bashar al-Assad's Last Stand: A Study of the Syrian Conflict's Final 11 Days in December 2024Three Winters in Exile: The Trump Chronicles 2021-2024Biden vs. Trump The 45th and 46th: A Tale of Two AmericasImran Khan's Political Journey: From Cricket to RevolutionPolitical Prophecies of Imran KhanFrom Captain to Khan: The Rise and Fall of Imran KhanThe Widening Political Gap Between Muslim Rulers And Their Masses In The 21st CenturyIbrahim Traoré

International Relations & Geopolitical Analysis

Can Russia Help Pakistan Grow?Why Nuclear-Armed Pakistan Has Not Fought a War Against Israel in the 2020s?A Textbook Of Foreign Policy AnalysisInternational Relations from a Pakistani PerspectiveThe UNO's Three Failures: Gaza, Kashmir, and Ukraine

Military Operations & Strategic Studies

27 Minutes That Nearly Started World War IIIOperation Bunyan um Marsoos2nd Edition:

Operation Rising Lion: Israel's Strike Against Iran's Nuclear ProgramOperation True Promise 3The Spider's Web: How Ukraine Rewrote the Rules of WarOperation Rising Lion 2025: Israel's Strike on Iran's Nuclear Program

Intelligence and Espionage Research Analysis

Behind The Veil Of Deception: Catherine Perez-Shakdam

Read more at https://naimtahirbaig.netlify.app.

www.ingramcontent.com/pod-product-compliance
Lightning Source LLC
LaVergne TN
LVHW080034210926
848481LV00011B/3184
* 9 7 9 8 2 3 0 1 0 6 4 4 9 *